Every door hides a story, discover them all at **onitsukatiger.com**

MADE OF JAPAN

Onitsuka Tiger™

Sneaker Tokyo vol.2
"Hiroshi Fujiwara"

前言

由《Shoes Master》所延伸出的附屬書籍《Sneaker Tokyo》。這次第二期將會以整整一冊來介紹東京球鞋文化的教父級人物「藤原浩」。

在音樂、時尚、藝術等領域皆有「當代首屈一指的酷獵研究者（Cool Hunter）」的美譽，藤原浩在球鞋的領域裡，將其獨特精準的眼光和敏銳的市場嗅覺發揮得淋漓盡致。從1980年代至1990年代，由他所穿過而引爆潮流的鞋款如「DUNK」、「COURT FORCE」、「AIR JORDAN」、「SUPER STAR」、「CAMPUS」、「JACK PURCELL」、「NORTH WAVE」等等不勝枚舉。進入2000年後，更透過他與NIKE長期友好的關係而發展出專屬於他個人的產品線HTM，推出了無數的經典名作鞋款。本書以近80頁的篇幅整理出藤原浩個人與球鞋的相關歷史脈絡，其他還有和他在公私兩面皆有深交的NIKE CEO Mark Parker等球鞋業界重要人物的對談訪問，更收錄了許多以他為軸心觀察東京球鞋文化歷史的專題報導。

很遺憾地，本書中所介紹的多款球鞋皆已絕版，在市面上已經找不到了。但是相信對球鞋愛好者來說，閱讀就有著相當高的樂趣，我們可以肯定本期相當具有資料保存價值。讓我們期待有些球鞋經過再次介紹後有了更新的評價，引發強大的復刻需求之後，能再次重新站上舞台。

Foreword

Sneaker Tokyo is a hardback spin-off from the Shoes Master magazine. This second volume in its entirety is a feature on Hiroshi Fujiwara, a man whose name has become synonymous with Tokyo's sneaker culture.

Hiroshi, who is reputed as the ultimate pursuer of all of "cool" in a range of occupations such as music, fashion and art, has over the years applied his sharp sense and phenomenal eye for quality to the world of sneakers. Throughout the 80's and 90's, countless models such as the Dunk, Court Force, Air Jordan, Superstar, Campus, Jack Purcell, and North Wave became popular in Japan, simply because Hiroshi was wearing them. Since 2000, he has produced a number of premier sneaker lines, such as HTM through his close relationship with Nike. This publication archives Hiroshi's historical connection to sneakers in just under 80 pages. You will also find interviews with other key figures from the industry, such as Nike CEO, Mark Parker (with whom Hiroshi associates with both on a private and business level), and a record of Hiroshi's past in correlation with the history of Tokyo's sneaker culture.

Regrettably many of the sneakers featured are no longer available in stores, but we are confident this publication may be enjoyed in full as a valuable documentation that takes another look again at sneakers of the past, making sneaker freaks everywhere wish that nostalgic pair will somehow make a comeback into the market.

Contents

HF Style
in 2010

NIKE_ALL COURT CANVAS, NIKE_HTM2 RUN BOOT LOW,
NIKE_AIR JORDAN 1 Retro, NIKE_TENNIS CLASSIC SB,
NIKE_TENNIS CLASSIC, NIKE_TENNIS CLASSIC V,
NIKE_BLAZER SB, NIKE×X-girl_BLAZER MID,
CONVERSE_JACK PURCELL, new balance_M1300

NIKE_ALL COURT CANVAS

[L] NIKE×X-girl_BLAZER MID [R] CONVERSE_JACK PURCELL

NIKE_HTM2 RUN BOOT LOW

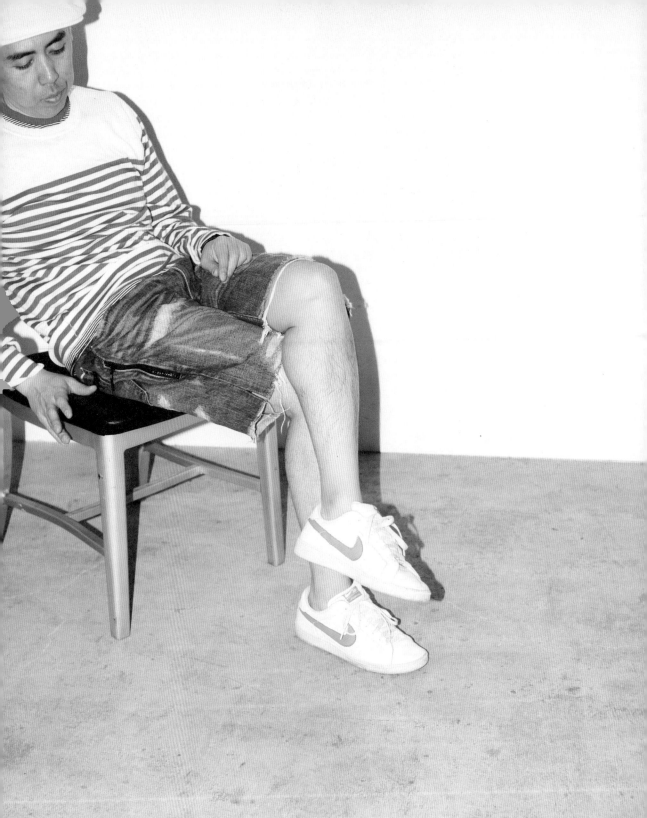

NIKE_TENNIS CLASSIC SB

NIKE_HTM2 RUN BOOT LOW

'NIKE_AIR JORDAN 1 Retro

[L-R] NIKE_TENNIS CLASSIC / NIKE_TENNIS CLASSIC SB / NIKE_BLAZER SB

NIKE_TENNIS CLASSIC V

NIKE×X-girl_BLAZER MID

new balance_M1300

Hiroshi Fujiwara's Sneaker Archive

藤原浩的球鞋記錄

從1980年代的「DUNK」、「AIR JORDAN」、「SUPER STAR」，
到1990年代「FOOTSCAPE」、「NORTHWAVE」、「Gravis」，
再到2000年代的「HTM」「visvim」、「TENNIS CLASSIC」，
我們把藤原浩所「選擇過的」、「穿過的」及「創造過的」約200雙球鞋，
完全整理記錄下來。

From the 80s, Dunk, Air Jordan and Superstar;
from the 90s, Footscape, Northwave and Gravis;
and from the past decade, HTM, visvim and Tennis Classic.
We have compiled an archive of around 200 pairs of the shoes
Hiroshi Fujiwara has chosen, worn and created.

NIKE
HTM

HTM是NIKE的一個特別的實驗企劃案，其名字是來自參與此計劃的3大人物（藤原浩、NIKE設計師Tinker Hatfield、NIKE CEO—Mark Parker）名字第一個字母縮寫組合而成。在最初期，這名字原本只是內部相關人士用來稱呼此實驗計劃的代號，但最後在商品正式推出之際卻直接用這個代號來命名了。HTM第一次向外界公開是在2002年的《Men's Non-No》雜誌裡的連載單元「A Little Knowledge」。之後，藤原浩又再推薦「WOVEN BOOTS」、「WOVEN」、「AIR MOC MID」等3款鞋款。接著，HTM在2004年推出充滿震撼話題性的「AIR FORCE 1」，2005年又推出籃球鞋「COURT FORCE」等鞋款。HTM的做法就是把已存在的鞋款重新配色處理過後，重新推出新的樣式，後來因為有許多鞋款大受歡迎進而發展出很多普通常賣款式。不過不管哪一款HTM都是限量生產（編入獨立編碼，各在1500～3000雙左右），而且也在限定的店舖販售，因此造成了球鞋狂熱者之間的超級爭奪戰，甚至在許多特別版都出現飆到天價的狀況。到了2009年，HTM加入了設計師Mark Smith，變化成新企劃案HTM2。

The NIKE HTM line is both unique and experimental. The HTM acronym represents the three creative talents who took part in the collaboration: Hiroshi Fujiwara, NIKE designer, Tinker Hatfield and NIKE CEO, Mark Parker. Originally HTM was simply a codename used by insiders, but the title remained unchanged when it hit the market. The line was first unveiled in 2002 in a series featured in Men's Non-No magazine, "A Little Knowledge", where AIR WOVEN BOOTS, AIR WOVEN and AIR MOC MID where introduced by Hiroshi Fujiwara himself. In the following years HTM came out with several more models, such as AIR FORCE 1 in 2004, and COURT FORCE in 2005. Some where new arrangements of existing designs, while HTM also announced the release of brand new models as well, many of which went on to become "inline" (a mainstream NIKE catalogued product). Sneaker freaks scrambled for this scarcely produced line (between 1500 and 3000 serial numbered pairs per model), sold only at exclusive stores fetching high prices on the street. In 2009, it was announced that HTM2, a new project featuring NIKE designer Mark Smith would be released.

HTM
AIR FORCE 1

HTM的「AIR FORCE 1」最大特色是使用高級材質，分別有兩款設計，第一款是採用高級皮革並縫上反白車線（之後也推出普通常賣款），而另一款則壓上鱷魚紋，同為2002年推出。

HTM AIR FORCE 1 is characterized by the use of premium material. The model comes in two designs, one featuring high quality leather with stitching and the other featuring a croc leather design. Released in 2002. (Later became in-line NIKE product.)

HTM
AIR WOVEN

雖然藤原浩也有參與初期「AIR WOVEN」的顏色配置，但另外HTM所推出的版本所運用顏色卻截然不同。這兩款是大膽採用多重色彩編織的「Rainbow」，2002年推出。

Hiroshi contributed to the color coordination of the original AIR WOVEN, as well as this HTM version of the model. These two models entitled RAINBOW feature an ambitious use of color. Released in 2002.

HTM
AIR WOVEN BOOT

2002年推出首批HTM系列中的款式。當時只有短筒版的「AIR WOVEN」，由藤原浩提出了新看法：「應該要有更時尚更容易搭配的中筒鞋款」，所以具體生產之後便得出此鞋款，材質為皮革及麂皮兩種。

Released in 2002 as part of HTM's first collection. Hiroshi Fujiwara came up with the idea to develop a mid-cut version of AIR WOVEN, designed to be easier to integrate with fashion. The model comes in both leather and suede.

HTM
AIR FOOTSCAPE WOVEN

「AIR FOOTSCAPE」的鞋底加上「AIR WOVEN」的
鞋身所組成的Hybrid混血款。在2005年HTM推出4款新
色鞋款時，以嶄新的販賣手法在NIKE JAPAN的官網透
過抽選方式販售。

AIR FOOTSCAPE woven is a hybrid model taking the sole
from the AIR FOOTSCAPE model and upper from AIR
WOVEN. These four colors were released from HTM in
2005, and sold thru the NIKE Japan website to randomly
selected consumers.

HTM
ZOOM MACROPUS

此鞋款特色在於看起來像普通休閒鞋般的獨特外形。
鞋底配有氣墊，穿上後十分舒適。在HTM第一次以實
驗的方式推出後，也成功地變成NIKE的常賣鞋款。

Characterized by its unique design. The Zoom Air sole
guarantees a comfortable fit. After an experimental
release under the HTM brand name, the model was also
released as an in-line NIKE product.

HTM
AIR PRESTO ROAM

採用伸縮性且柔軟的鞋身材質，這是休閒鞋「AIR PRESTO」的高筒版。材質改用麂皮、顏色以棕色系配色帶出沈穩的感覺。2002年推出。

High-cut version of AIR PRESTO. This model features a soft stretch upper. As opposed to the original, AIR PRESTO ROAM is made with a soft brown suede material, and "classic" in style. Released in 2002.

HTM
AIR MOC MID

因為圓滾滾的鞋形而獲得超高人氣的「AIR MOC」中筒版，2002年推出。為了增加穿脫的便利性，在側面特別加入富彈性的材質。

Mid-cut version of the AIR MOC, released in 2002. Popular for it's unique rounded form. Its rubber sides make the shoe easy to slip on and off.

Archives

NIKE

HTM

HTM
COURT FORCE

藤原浩在1980年代玩滑板時經常穿著的籃球鞋「COURT FORCE」。這款他十分喜愛且充滿回憶的球鞋在2005年由HTM推出復刻版本。

When Hiroshi Fujiwara was skateboarding in the 1980s, he was inseparable from his COURT FORCE basketball shoes. This emotional involvement brought him to revive the model under HTM in 2005.

HTM2
RUNBOOT TZ

新加入的設計師Mark Smith所推出的2009 HTM2第一款，鞋底採用NIKE FREE 7.0，鞋身則採用無鞋帶充滿未來感的設計。

First model from HTM2, a new collection announced in 2009 this time also featuring Nike designer Mark Smith. Neo-futuristic design featuring NIKE FREE 7.0 technology sole.

HTM
AIR TERRA HUMARA

戶外專用球鞋「AIR TERRA HUMARA」的HTM版本，採用高級皮革及縫上反白的中線。和之前推出的「AIR FORCE 1」有著相同的手法，於2002年推出。

HTM version of AIR TERRA HUMARA, a sneaker designed for the outdoors. This shoe featured high quality leather with stitching in a new styling similar to the previously released AIR FORCE 1. Released in 2002.

NIKE
DUNK

「DUNK」是1985年誕生的經典籃球鞋，但是在1980年代後期，藤原浩穿上它卻是應用在滑板運動方面。他對於「DUNK」有以下看法：「鞋底的抓地力十分強，和其它鞋款相比起來中底較薄（也就是能獲得恰到好處的著地感），所以非常適合用來玩滑板呢。」他最喜歡的配色是「深藍×黃」，而這個配色在「DUNK」中也是最為少見且價錢極高的一款，藤原浩卻徹底地把它應用在滑板運動方面，完全不見他感到可惜的神色，甚至現在已穿到鞋子都快破爛了。設計方面，「DUNK」的結構及外觀跟同期推出的「AIR JORDAN 1」非常相似，藤原浩回憶說：「那時我以為這是AIR JORDAN的不同配色所以才買下『DUNK』的。」進入1990年代後，「DUNK」成為稀少球鞋的代名詞，在不少二手市場也以超高價在流通發售，從誕生後14年的1999年，終於有了令人期盼已久的復刻版上市。之後，更成為代表NIKE的基本鞋款之一，近年不斷推出許多不同的配色及別注版本。藤原浩本人平時也十分酷愛穿它，後來也進一步和NIKE合作，推出了圓點等不同款式的設計。

The DUNK is a basketball shoe that was originally developed in 1980. In the late 1980s, Hiroshi Fujiwara wore the shoes not for basketball, but for skateboarding. Hiroshi claims the outer sole has a high-grip, and compared to other sneakers the mid-sole is thinner (creating a better sense of contact with the ground), so the shoes were perfect for skateboarding. His favorite color was "navy and yellow". This particular color was a rarity, but Hiroshi abused the sneakers with his skateboarding wearing them out as long as he could. As a design, the shoes show a significant resemblance to "AIR JORDAN 1", a model that was released around the same time. Hiroshi recalls that he bought the shoes thinking they were Jordan's that had come out in a new color. In the 1990s the DUNK came to be defined as a rare sneaker and entered the vintage market with a substantial price tag. In 1999, fourteen years after it was first released, the model made a long-awaited comeback. Since then the DUNK has become one of NIKE's standard models, spurring a great number of variations upon the original design. Hiroshi has contributed to versions such as the POLKA DOT through his collaborations with NIKE, and also enjoys wearing the shoes for his everyday life.

1980年後期至1990年代前期，藤原浩在玩滑板時所穿著的深藍×黃「DUNK」。

DUNK - Hiroshi wore for skateboarding from late 80s to early 90s.

鞋身採用咖啡色高級皮革的高價款式，2002年推出。

Quality Brown Leather Premium Model (2002)

鞋身採用麂皮和尼龍材質組合而成的「UNDEFEATED」特別版，2003年推出。

Custom designed model by UNDEFEATED made from suede and nylon (2003)

藤原浩參與擔綱設計，通稱「圓點」（Polka Dot），
2006年推出。
POLKA DOT co-designed by Hiroshi Fujiwara (2006)

NIKE和木村KAELA合作的鞋款，鞋舌註明KAELA字
標，2006年推出。
Collaboration model from NIKE and Kaera Kimura (2006)

NIKE
AIR FORCE 1

1982年登場的經典「AIR FORCE 1」,是中底內首次加入AIR氣墊的藍球鞋。和3年後推出的另一藍球鞋「DUNK」成為兩大最受歡迎的NIKE代表鞋款,至今推出的配色多不勝數,同樣受到當今所有球鞋愛好及收藏者的喜愛。設計特點方面,「AIR FORCE 1」相比其他藍球鞋來說,鞋底明顯較厚,所以被藤原浩批評為「不適合用來玩滑板的鞋款」。在2002年發售的HTM版本及2009年發售的25周年紀念鞋款中,有不少款式都是由藤原浩本人親自負責擔綱設計。另外,也有很多其他款式是他本人十分喜歡而經常穿著的。整體來說,他所選的所有「AIR FORCE 1」鞋款都有一個共同特色,就是全以簡單的配色搭配高級的材質。

The AIR FORCE 1 was released in 1982, features Air technology in its mid-sole. It was the first ever basketball shoe with air bag from Nike. The equally popular DUNK came on to the scene three years later and became one of NIKE's standard models and to this day is loved by sneaker freaks worldwide. As a design, the sole is thicker than other basketball shoes. Hiroshi comments this design aspect makes it inappropriate for skateboarding. As well as wearing the shoes on a daily basis, Hiroshi was involved both in the production of the HTM version released in 2002, and the 25th anniversary edition in 2009. Both shoes feature simple coloring and quality leather.

2002年為了「W+K TOKYO」廣告公司,藤原浩設計並製作了這款特別的非賣品。鞋面用上還沒有出生的小牛皮革,鞋跟左右二處則把「東京」2個字以刺繡方式作點綴。

In 2002, Hiroshi Fujiwara designed and produced this model for advertising company, Wieden + Kennedy Tokyo. Not for sale. Unborn calf hide has been used for the upper. "Tokyo" has been embroidered in Japanese characters across the heel.

由NIKE贈送給藤原浩的超級持別款。鞋身完全採用純
鱷魚皮，相當豪華的設計。非賣品。

Shoes presented to Hiroshi Fujiwara by NIKE. Authentic
crocodile leather upper. Not for sale.

「AIR FORCE 1」25周年紀念款，於2007年推出。此
為藤原浩所提出的配色。

25th anniversary AIR FORCE 1 model released in
2007. Color coordinated by Hiroshi Fujiwara.

Louis Vuitton的貴賓室「CELUX」(2008年結束) 所訂製的鞋款。在2006年, 此貴賓室4周年紀念時曾經推出過「NIKEiD Studio」。

This is a custom order shoes at Louis Vuitton members-only salon "Celux". (Closed in 2008) In 2006, NIKE iD Studio was incorporated into the same salon for a limited period.

Logo「SWOOSH」上加上GUCCI的圖案, 這是某店所製作的非官方鞋款, 是「自行合作」的做法。

Gucci pattern embossed on SWOOSH shoes. Unreleased model produced by a shop.

這3雙加上閃電圖形的是fragment design設計的版本。

These are special labeled shoes by FRAGMENT DESIGN with the trademark "the thunder".

NIKE
AIR JORDAN

眾所皆知，這款是史上最強NBA籃球選手Michael Jordan的同名鞋款「AIR JORDAN 1」。1985年推出以來，以一年一雙的速度推出新的款式，在球鞋愛好者間也成為了必然蒐集的名品，相當具有人氣。第3代以後由HTM其中一位成員Tinker Hatfield負責擔綱設計。隨著一代一代的演進下去，無論外形或功能性兩方面都愈顯卓越及進步。歷代中除了某些鞋款外，藤原浩就有以下的感受：「我不是太喜歡第5代以後的鞋款設計。」

This is the most famous signature shoes by the famous basketball player of all time, Michael Jordan. A new model from the series has been released every year since its birth in 1985. AJ became a popular collectors' item among sneaker freaks. Tinker Hatfield, also a designer for HTM, has contributed to the design of the series since the third model. Over the years, the design and technology of the shoe has advanced, but Hiroshi comments apart from the odd exception, he is not a fan of models released in the series after AJV.

這是第1代藍×黑的珍貴絕版名品，是藤原浩口中「AIR JORDAN系列中最喜歡的鞋款，也是最喜歡的顏色」。

Deadstock of AJ1 that is my favorite sneaker, Hiroshi said. He thinks AJ1 has the best design and color way in its series.

第4代粉藍×白的原型，1989年推出。

Original IV
(1989)

第4代的「UNDEFEATED」之特別訂製款，鞋身換上
軍事味道的配色，2005年推出。

Custom-designed UNDEFEATED version of IV.
(2005)

第11代的短筒原型，1996年推出。

Original XI Low
(1996)

第11代的原型，1995年推出。

Original XI
(1996)

NIKE
COURT FORCE

1982年「AIR FORCE 1」推出後，許多冠上了「FORCE」名稱的鞋款陸續推出，「COURT FORCE」就是在這股潮流中於1987年推出的籃球鞋。「COURT FORCE」和「DUNK」一樣，是藤原浩在滑板時十分愛用的鞋款。「當時在日本還買不到『COURT FORCE』，我是在紐約買的。我在當地看到滑板選手都穿著它玩滑板，感覺實在太酷了！」藤原浩回憶說。當時知名的滑板玩家如Eric Dressen及Jeff Kendall也是穿著「COURT FORCE」上場的。在鞋款推出20年後的2005年，由於藤原浩的要求，所以透過HTM這個實驗的計劃，達成的初次復活的願望，當時推出由「HECTIC」特別訂製的配色版本，之後也進一步發展成普通常賣款式，和「DUNK」、「AIR FORCE 1」並列為NIKE代表鞋款並持續熱賣當中。

Since the birth of the AIR FORCE 1 in 1982, several successor models have been released under the name "Force". The COURT FORCE is one such model which appeared in 1987. As with the DUNK, Hiroshi wore the COURT FORCE for skate boarding. "At that time, you couldn't buy the shoes in Japan, so I bought them in New York after I saw how good they looked on the professional skate boarders over there wearing them." As Hiroshi comments, professional skateboarders at the time such as Eric Dressen and Jeff Kendall were often spotted wearing COURT FORCE when they were skateboarding. 20 years after the birth of COURT FORCE, Hiroshi revived the shoe under the HTM line in 2005, with a model called HECTIC. Following that the model became in-line and joined the DUNK and AIR FORCE 1 as standard NIKE models.

1980年後期至1990年代前期，藤原浩在玩滑板時所穿著的黑×白「COURT FORCE」。
COURT FORCE shoes worn by Hiroshi for skateboarding from late 1980s to early 1990s.

和左頁屬同款但配色不同，同樣是藤原浩在滑板時所穿的「COURT FORCE」。

Same shoes as left in different colorway. COURT FORCE shoes worn by Hiroshi when he was skateboarding.

2006年由街頭品牌「HECTIC」推出的特別配色版本。

HECTIC releasedin2006.Original promotional colorway.

NIKE
TERMINATOR

和「DUNK」、「AIR JORDAN 1」同年於1985年推出的經典籃球鞋「TERMINATOR」。在鞋跟位置大大的「NIKE」字樣（通稱「BIG NIKE」）是外觀設計上的一大特點。根據藤原浩的提議，2009年推出的別注鞋款就把「NIKE」字樣大膽改為「NOISE」，同年所推出的「NOISE Tee」也就是這個點子的來源。此外，此款球鞋更採用了Himalaya WP高級皮革，運用黑×銀的配色令此鞋添上了未來科技的高級感覺，品質十分講究。另外，中底額外配備ZOOM AIR氣墊以提供緩震功能，大大增強了舒適度。除了外觀作出變化外，也同時提高了功能性的要求。

The TERMINATOR debuted in 1985, the same year as the DUNK and AIR JORDAN 1. One of the model's main design features is the large "NIKE" motif at the heel. Hiroshi came up with the idea to change the motif from "NIKE to "NOISE" when it was re-released in 2009. The idea originated from the NOISE Tee line released in the same year. The model features Himalaya WP premium quality leather, and a Zoom Air sole that works to cushion the foot. The shoe is enhanced from both a design and technology perspective.

黑色款是2009年推出，白色則是日本末發售鞋款。

Black released in 2009.
White not sold in Japan.

NIKE
DYNASTY

1985年誕生的藍球鞋「DYNASTY」，它的命運和同年誕生的「DUNK」、「TERMINATOR」不同，它並沒有成為主流鞋款，但卻被部分球鞋愛好者稱為「隱藏版名作」而獲得不少的支持。設計方面，和同期推出的「VANDAL」等同樣地搭配了踝部的魔鬼氈保護帶，是其一大特色。雖然藤原浩說：「這個時期幾乎大家都推出差不多的鞋款，誰是誰都分不清了。」但經過20年，如今仍有許多人將它視為基本愛用鞋款，也許「1985」這個年份是史上罕見地同時出現許多經典好鞋的年份吧！此外，「DYNASTY」於2009年成功推出復刻，2010年也將陸續推出新的配色。

Debuted in 1985, the DYNASTY is not as major as the DUNK and TERMINATOR lines that appeared in the same year, but some fans favor the line as a "hidden masterpiece". From a design perspective, similar to the VANDAL model released around the same time, the ankle strap is one of the main design focal points of the shoe. Hiroshi comments, "The shoes coming out at the time all shared similar designs to the point it was hard to tell the difference.", but bearing in mind the same models are still loved by many in the present more than twenty years later, it can also be said that 1985 was a prominent year in the history of the sneaker. The DYNASTY was re-released in 2009, with many new colors arriving in 2010.

1985年絕版品，能夠完美如新保存至今十分罕見。

Deadstock from 1985. Original DYNASTY shoes kept in such good condition considered to be extremely rare.

NIKE
ORCA PACK

在不同鞋款上運用相同的配色並化成系列一併推出，在球鞋業界被稱之為「PACK」的做法，近年可說十分流行。由藤原浩所操刀的「ORCA PACK」正是其一出色的例子，「ORCA」在英語裡有「殺人鯨」的意思，而「ORCA PACK」指的就是在4款不同的鞋款上運用黑白色搭配出有如殺人鯨般的簡潔感覺。

"Pack" is industry lingo for a design concept that means to apply the same coloring and patterns to a number of different models. In the ORCA PACK series, "Orca" or the killer whale has been used as a color inspiration applied to four different models.

上至下「DUNK HI」、「DUNK LOW」、「AIR FORCE 1」、「AIR TRANER 1」，4款皆在日本國內僅兩店舖限定販售，2004年推出。

From the top, DUNK HIGH, DUNK LOW, AIR FORCE 1, AIR TRAINER 1. All four designs were released in 2004. Sold at only two stores in Japan.

NIKE
MONOTONE COLLECTION

2001年推出的「MONOTONE COLLECTION」是NIKE JAPAN首度和藤原浩合作的鞋款，如同系列名稱，就是將已存在的鞋款再度以黑白色調的方式重新推出。被選入此系列的鞋款為當時十分流行的「AIR TERRA HUMARA」、「AIR ZOOM SEISMIC」及「AIR MAX 120」此3款。這個構想不但十分獨特，而且這種「重新以相同概念系列化」的包裝概念在當時的日本是相當罕見的做法。

Released in 2001, the MONOTONE COLLECTION was the first collaboration between NIKE Japan and Hiroshi Fujiwara. The concept was literally to re-produce existing models in monotone. Three models were selected for the collection: AIR TERRA HUMARA, AIR ZOOM SEISMIC and AIR MAX 120. The idea was not only unique but also the collection came from Japanese talent which was unusual at that time.

「AIR TERRA HUMARA」、「AIR ZOOM SEISMIC」、「AIR MAX 120」此3款雖然都以黑白色調呈現出來，但卻在不經意的細節上用了綠色點綴。

AIR TERRA HUMARA, AIR ZOOM SEISMIC, AIR MAX 120 Each model consists of a monotone base with green accents.

NIKE
AIR FOOTSCAPE

1995年誕生的「FOOTSCAPE」，藤原浩稱之為「給他很大衝擊的一雙球鞋」。「左右不對稱的鞋面，適合日本人腳形的鞋身設計，略為偏圓的鞋頭……所呈現出來的全部都是嶄新的點子，我自己非常喜歡，經常穿著之外也會不斷尋找新的配色。」2009年由fragment design操刀推出了全新的配色，鞋面完全運用網狀材質，鞋頭則採用了麂皮，這是藤原浩對1995年原型的配色作出致敬之意思。

The FOOTSCAPE was launched in 1995. Hiroshi said that it was one of the most sensational sneaker series for him. "Asymmetrical upper, wide width to fit Japanese feet and rounded toes.... Everything about the shoe was new to me. The shoes were one of my favorites and I was often on the lookout for new colours." In 2009, a FRAGMENT DESIGN colored version was released. It was Hiroshi's homage to the original model. Features all mesh upper and gray suede toe.

2009年推出由fragment design所操刀的鞋款，共4色。
FRAGMENT DESIGN model released in 2009. Sold in four colors.

FOOTSCAPE最經典的配色，也是1995年面世時的原色版本。

FOOTSCAPE's most well-known color, debuted in 1995.

NIKE
AIR WOVEN

2000年誕生的「AIR WOVEN」,靈感是來自在美國奧勒岡州洞窟內所發現達10,000年之久的舊布,進而以此布的編織方法為概念開發而使用在球鞋上,如同鞋款名稱「WOVEN」(編織)的作法,展現出輕巧、高透氣、高彈性等特色。此外,在普通的球鞋生產過程中,材料經過切割後會產生廢料,但「WOVEN」卻不是把材料「切割」、「縫死」,而是以「編織」方式成形,所以幾乎不會產生廢料,是其一大特點。自從由藤原浩擔綱配色推出第一代後,陸續由HTM推出之後的系列,常賣款中也不定期會推出新色。後來,NIKE還利用這種編織方法推出其他鞋款如「AIR PRESTO WOVEN」及「AIR FOOTSCAPE WOVEN」等。

The AIR WOVEN first released in 2000. These shoes were inspired by a ten thousand year old cloth, discovered in a cave in Oregon. The upper is literally made with woven material, which is light, breathable and flexible. Generally waste is produced from left over sneaker material, but woven models are made by weaving material, instead of cutting or sewing, so produces substantially less waste. After the first collection Hiroshi contributed to the coloring of the series. This model was released from HTM and has also been known to feature in NIKE's inline catalogues on the odd occasion. AIR PRESTO WOVEN, AIR FOOTSCAPE WOVEN and other models using the same material were also released.

「AIR WOVEN」原始鞋款，其配色由藤原浩操刀。

The first collection of AIR WOVEN. Colored by
Hiroshi Fujiwara.

NIKE
AIR FOOTSCAPE WOVEN

把兩種不同的結構合而為一所生產出來的鞋款，稱之為「Hybrid」，這樣的手法在球鞋業界也時不時地被拿來運用。由「AIR WOVEN」和「AIR FOOTSCAPE」所融合出來的「AIR FOOTSCAPE WOVEN」就其中代表之一，這就是藤原浩曾說過「把兩款喜歡的鞋款合而為一」的新鮮做法。2005年由HTM以實驗性質的方式推出後，2006年再推出圓點圖樣以及靴型款，全部在指定的店舖內以數量限定的方式販賣，之後也進而擴展至常賣款系列之中。

This model is referred to as a "hybrid", a method often used in the sneaker industry. A good example is AIR FOOTSCAPE WOVEN, which is a combination of two different models: The AIR WOVEN and AIR FOOTSCAPE. These two models are favorites of Hiroshi. This model was tentatively released under the HTM line in 2005. After that, a polka-dot pattern design and a boots version of the model were released in limited quantities at limited stores. Later production of the model expanded when it became an inline NIKE product.

由藤原浩所提案的圓點圖樣款，2006年在全球主要都市的指定店舖內販售。

Polka-dot pattern model conceived by Hiroshi. This model was sold only at exclusive stores in major cities around the world in 2006.

高筒的靴型款。這也是2006年在全球主要都市的指定
店舖內販售。

High-cut boots type, sold only at exclusive stores in major
cities around the world in 2006.

Archives

NIKE

TENNIS CLASSIC

NIKE
TENNIS CLASSIC

2010年的今天，東京捲起了一股從未有過的網球鞋風潮。由於藤原浩在2007年從NIKEiD訂製了「TENNIS CLASSIC」而開始愛上此鞋款並經常穿著，之後更與NIKE進行了多次合作推出許多不同的版本。其實，「TENNIS CLASSIC」是超過30年歷史於1978年所推出的經典鞋款，外形和厚重的藍球鞋不同，呈現出簡潔且俐落的線條，或許正因如此，和現今的街頭氛圍有著相同的取向吧。藤原浩說：「adidas有『STAN SMITH』這樣定番的網球鞋款，NIKE卻沒有。我認為這款『TENNIS CLASSIC』若也能有相同地位就好了」。除了NIKEiD之外，由藤原浩所擔綱設計的款式，所有的Swoosh Logo都以打洞方式處理，這樣的展現手法是從1973年所推出的網球名鞋「WIMBLEDON」所獲得的靈感。

Now in 2010, tennis shoes are more popular than ever in Tokyo. Hiroshi also enjoys wearing his pair of TENNIS CLASSIC's since he ordered them online at NIKEiD. Since then he has worked on numerous models in this line through his collaboration with NIKE. The model TENNIS CLASSIC, a milestone in sneaker history, was introduced in 1978 more than 30 years ago. It has a simple and sharp shape; a radical change from the fat shape of the more popular basketball shoes of the time, yet resonant of today's street atmosphere. Hiroshi says, "Adidas has Stan Smith which is their popular tennis shoe, something NIKE doesn't have. I'm hoping the TENNIS CLASSIC model will become that kind of shoe in the NIKE repertoire. All the TENNIS CLASSIC directed by Hiroshi except NIKEiD carries perforated SWOOSH logo. The idea for this style comes from the classic tennis shoe, WIMBLEDON which appeared in 1973.

為了2008年5月所發生的四川大地震而設計的義賣鞋款，鞋舌部分印有「四川加油」的文字。

Shoes made for the charity auction in support of the reconstruction of Sichuan where a major earthquake occurred in May 2008. The message, "四川加油" Cheer-up is written on the tongue.

2007年藤原浩向NIKEiD訂製的鞋款，鞋舌及鞋跟為金色，滾邊則為紅色，是簡單又搶眼的配色。

Hiroshi ordered these custom-order shoes online at NIKEiD. A simple combination of gold for the tongue and heel, and red lining.

由藤原浩及SOPH的清永浩又共同品牌uniform experiment 與NIKE合作推出的鞋款，2009年推出。

Collaboration model by UNIFORM EXPERIMENT (joint brand of Hiroshi Fujiwara and Hirofumi Kiyonaga from SOPH) and NIKE. Released in 2009.

2008年10月，原宿「NSW STORE」開幕時推出的紀念鞋款。

Anniversary model which was produced for the NSW STORE opening in Harajuku in October 2008.

NIKE
MATCH

1973年誕生的網球鞋「MATCH」。2009年由
UNDERCOVER和fragment design以共同名義重
新復刻。藤原浩表示「這雙鞋的鞋頭較薄，騎自
行車時穿剛好。」設計方面，在鞋側的SWOOSH
Logo這麼小是有原因的，因為當時的網球比賽服
裝有嚴格的規定，過大的SWOOSH是被禁止的。

The MATCH tennis shoe was launched in 1973.
First reproduced jointly by UNDERCOVER and
FRAGMENT DESIGN in 2009. Hiroshi says, "This
shoe has a slim toe and is good for cycling." The
small side SWOOSH logo is characteristic of this
design. It's said that there was a strict tennis dress
code at the time so a lavish SWOOSH logo would
not have been regarded appropriate.

鞋底外側會印有UNDERCOVER及fragment design的
Logo，2009年推出。

Both UNDERCOVER and FRAGMENT DESIGN logos are
printed on the outer sole. Released in 2009.

NIKE
ALL COURT

1975年推出的「ALL COURT」，顧名思義正是
多功能鞋的先驅，為了各種不同類型的運動所設
計的鞋款。2009年和A.P.C.合作推出帆布款式，
同年fragment design又推出了高級材質的高價
款。鞋頭偏圓，鞋身側面以打洞方式呈現出大大
的Swoosh Logo，都是其設計上的特點。

ALL COURT was released in 1975. Literally an "all-
round" shoe, and pioneer for the CROSS TRAINER.
In 2009, a canvas shoe model was reproduced in
collaboration with A.P.C.. In the same year, a
premium version of this model was released,
produced by FRAGMENT DESIGN. The rounded toe
and SWOOSH logo punched on the side, forms the
focal point of this design.

fragment design所推出的高價版，採用如同雷射玻璃殼具有光
澤的皮革，2009年推出。

Premium model designed by FRAGMENT DESIGN. Using shiny
leather like glass leather. Released in 2009.

NIKE
SOCK DART

「原本打算由HTM來推出」的「SOCK DART」，是以結合「襪子」和「球鞋」的概念所研發出來的野心之作。鞋款具有優異伸縮性的鞋面，有如襪子合腳般舒適，腳背上的帶子更具有調整鬆緊的功能，為鞋子和腳部之間帶來最大的穩定性。「穿著起來十分舒適，而且鞋身的配色也很棒，很適合夏天，只可惜當時生產的數量相當有限，很難才可買到呢……能夠做出這種具有高度挑戰性的鞋款，就是NIKE的高超本領。」

The"SOCK DART was originally scheduled to be released under the HTM line". This is an ambitious work, which was developed based upon the concept of "socks and sneaker hybrid". The super stretchy upper and high support with adjustable strap on the instep allows the shoes to fit your feet like socks. "It's comfortable to wear and the colors of the upper have come out well. I'll be wearing them this summer. Only a limited number of shoes have been made, so they're difficult to get your hands on. One of the great things about NIKE is that they have the capacity to produce this kind of challenging model."

2004年採限量發行方式推出，共6色可供選擇。
Released in a limited quantity in 2004 in 6 colours.

NIKE
AIR PRESTO

2000年推出的傑作，概念是運用像T-SHIRT般的S/M/L/XL來劃分鞋子的尺碼，引發出這個前所未有的話題鞋款便是「AIR PRESTO」。它的定位是「運動過後穿著的舒適鞋款」，鞋面的特點是採用像襪子般伸縮性極佳的網狀材質。雖然鞋身是附有鞋帶的設計，但卻能夠輕易穿上，步行時十分舒服。下方相片中的「格子旗」版本是藤原浩訂購的NIKEiD限定鞋款。

Released in 2000. The AIR PRESTO attracted attention because of the sizes the shoe comes in: S/M/L/XL just like T-shirts. This model was developed as a "comfortable shoe for after exercising" and has a stretchy mesh upper. Although this is a laced shoe, it is very easy to put on and off. The photographed checkered flag pattern model was only available on NIKEiD. "I ordered this model in a hurry because I knew NIKE wouldn't be making any more of the pattern."

格紋圖案是2002年僅限NIKEiD所推出的款式。「當時，聽說這款快要賣完了，我趕快買了一雙。」

This checker black pattern was only ever available from NIKEiD in 2002. "I hurried to order this pattern, because I heard it was going to sell out quickly."

HELLO KITTY 30周年時由藤原浩所設計的特別版鞋款。是2004年為宣傳用而製作出來的非賣品。

A special model produced by Hiroshi to celebrate HELLO KITTY's 30th anniversary in 2004. This model was for promotion purposes only and not for sale.

左邊是NIKE特別送給藤原浩的特別版本，右邊則是電影《SEX AND THE CITY》宣傳用的鞋款。

Shoes on the left are a special edition model, gifted to Hiroshi by Nike. It is thought the shoes on the right were made for the promotion of the movie, SEX AND THE CITY.

NIKE
AIR MAX

回顧藤原浩過往的球鞋歷史，似乎沒有穿過太多慢跑鞋的跡象。1987年誕生，配合當時最新科技的「AIR MAX」，他似乎好像不怎麼在乎，另外對於造成社會現象大受歡迎的「AIR MAX 95」，他甚至曾經明確表示「不太喜歡」（他買的第一雙是AIR MAX 98）。不過話雖如此，藤原浩對「AIR MAX」有以下的感受：「只要是搭載新避震功能的鞋款，一推出我還是想要試試看穿起來的感覺。」搭配360度VISIBLE AIR的「AIR MAX 360」，據說他很喜歡穿起來的感覺。

Looking back on the sneaker history of Hiroshi Fujiwara, there were only a few times he wore running shoes. The AIR MAX has always utilized the latest technology of the day since it was first launched in 1987, although Hiroshi displays little interest in the series. As for AIR MAX 95 which became a social phenomenon, Hiroshi comments, "I didn't like them much.". (The first pair of AIR MAX he bought was AIR MAX 96.) Having said that when new cushioning technology comes out, there is always that little part of Hiroshi that is curious to test it out. He particularly likes AIR MAX 360, which features a visible air unit that is visible from a 360 degree angle.

2006年推出的「AIR MAX 360」，直接把AIR氣墊充當鞋底的終極AIR MAX。

The AIR MAX 360 was the ultimate AIR MAX with a sole made of air itself. Released in 2006.

這是與左頁不同色款的「AIR MAX 360」。
AIR MAX 360 in different colours.

首度搭配VISIBLE AIR的「AIR MAX 98」。
AIR MAX 98 featuring first full -length visible air unit.

「AIR MAX 90」的圓點圖樣版本。這是倫敦的球鞋店
「FOOT PATROL」的特別訂製鞋款。
Dot pattern model of AIR MAX 90. Custom model from
FOOTPATROL in London.

NIKE
SFB

「SFB」是「SPECIAL FORCE BOOTS」的簡稱，原來是為了美國陸軍所量身訂作的戰鬥專用靴。外層鞋底的抓地性高，中層鞋底搭配了避震性優異的材質，在極端嚴苛的岩石地面上也能輕鬆地行走。此外，它有著相比傳統靴子輕量的材質，不僅限於戶外，也很適合在平日穿著。

The SFB, which stands for SPECIAL FORCE BOOTS, are real combat boots originally designed for the US Army. They feature a high-grip outsole and high-cushion mid-sole designed for walking over boulders and rocks. May be used for day to day city wear as well as the outdoors, as they are so light, it hardly feels like you are wearing boots.

紫色是藤原浩所提案的顏色，中筒靴則是uniform experiment的特別鞋款，全系列都附有穿脫容易的拉鏈設計，皆為2009年推出。

Purple color model designed by Hiroshi. The mid cut customized model is designed by UNIFORM EXPERIMENT. All models have a zipper which makes it easy to on and off. Released in 2009.

NIKE
P-ROD2

在街頭擁有超人氣的職業滑板手Paul Rodriguez的同名鞋款「NIKE PAUL RODRIGUEZ 2 ZOOM AIR」，通稱「P-ROD」。由Tinker Hatfield、Sunday Voltega等知名的NIKE VIP所經手設計出特別配色後，再由藤原浩經手的便是現在這款，其名稱就是「FUJI ROD」，在鞋跟及鞋舌部位所看見的彩虹配色，靈感是來自眾所皆知的知名電腦品牌以前的Logo。

NIKE Paul Rodriges 2 Zoom Air, commonly known as the P-ROD 2, is pro skater Paul Rodorigas's signature model, enjoying considerable popularity on the street. After special models designed by NIKE's VIPs such as Tinker Hatfield and Sandy Bodecker were released, Hiroshi Fujiwara worked on this special model, FUJI ROD. The rainbow coloring on the heal and tongue is inspired by the logo of the world famous computer company.

2008年推出，僅限某些店舖販售。

This model was only sold at exclusive stores in 2008.

NIKE
TWENTYTWOSEVEN

2001年所推出的「TWENTYTWOSEVEN」，是一雙多功能運動鞋，分別有短筒及中筒，然後再分無鞋帶及寬式鞋帶這兩個版本。藤原浩說：「看起來就像機器人的腳。」這樣的設計對於運動來說是太過獨特了。雖然藤原浩本人曾一度對它熱衷而經常穿著，在當時的街頭也獲得一定程度的支持，但這樣的奇特的外觀，終究無法變成定番而導致它短暫的生命。

TWENTYTWOSEVEN, released in 2001, originally designed for cross training. Low cut and middle cut, shoe lace and strap types were released. The design is extremely unique for training shoes, as Hitoshi describes "they look like the feet of a robot". This model was his favorite for a while and also received a certain amount of popularity in the street. The popularity didn't last long though, probably due to the shoe's somewhat overwhelmingly unique design.

NIKE
CROSS TRAINER 2

2000年所推出的「CROSS TRAINER 2」如同鞋款名稱，是一雙多功能運動鞋。當時，藤原浩在雜誌《Men's Non-No》上的連載單元「A Little Knowledge」裡大力地推薦這雙球鞋，給了它「做到了多功能運動鞋最難作出變化的變化」這樣的高評價。的確，在鞋面設計和鞋帶的配置都有著嶄新的突破。

As the name suggests the NIKE CROSS TRAINER 2 was also designed for cross training. Hiroshi favorably introduced the model in his magazine series, "A Little Knowledge" in MENS NON-NO, commenting that "the shoe surpasses the category of sneakers". The cutback design on the upper and the position of the swoosh express true originality.

NIKE
AIR TIEMPO

「AIR TIEMPO」是針對室內足球等類型的運動所研發的鞋款。雖然藤原浩本人是不參與足球類運動的，但卻十分喜歡這雙鞋，原因是「這雙鞋鞋頭偏薄，很適合騎自行車時穿著。」這兩雙皆為藤原浩在NIKEiD所訂製的鞋款，鞋身側面有著「FRAGMENT」的文字刺繡標記。

The AIR TIEMPO is designed for training for indoor soccer and futsal. Hiroshi doesn't play soccer or futsal, but he was fond of this model because "the slender design of the toe is perfect for cycling". These are custom made by NIKEiD. On the side of the upper, the word "FRAGMENT" is embroidered.

NIKE
AIR ZOOM FLIGHT 95

在高科技球鞋風潮極盛的年代1995年推出的籃球鞋。側面
有如氣球般巨大的嶄新配件設計引起極大的話題，不僅在
籃球場甚至在街頭都極具人氣。鞋如其名，搭配ZOOM AIR
氣墊，具有適度的避震效果。2009年也推出復刻版。

Basketball shoes released at the peak of the high-tech sneaker
boom in 1995. The huge eyes on the side and the carbon like
texture caused much hype. The shoes became popular not only in
basketball circles, but also as street shoes. As its name suggests,
it features a Zoom Air unit and provides moderate cushioning.
Reproduced in 2009.

NIKE
AIR TRAINER 1

AIR TRAINER是為了運動訓練時保護腳部專用的系列。自1987
年「AIR TRAINER 1」面世，由於大受歡迎而成為人氣系列。此
鞋款最著名的設計細節就是魔鬼氈搭帶，除了有裝飾用途外，也
有功能性的意義。「能夠確實固定腳步，我在健身房內經常穿
它」。相片為加入蛇紋的「蟒蛇」系列，在2003年推出。

The AIR TRAINER series was developed as a shoe to protect feet in
training. The popularity and success of the first model of the AIR TRAINER
released in 1987, spurred a number of variations of the shoe in subsequent
years. The velcro strap which is a design icon for this model is not just an
ornament but clearly has a functional meaning. "The strap supports your
ankle securely. I used to wear these at the gym." The PYTHON PACK
model is photographed. Released in 2003.

NIKE
CROGPOSITE

籃球鞋「FORMPOSITE」在1997年以聚氨酯（Polyurethane）
材質鞋面直接包覆至中底的前衛設計引爆超人氣話題，後
來再以此鞋款大膽地改以拖鞋式設計，變成現在看到的
「CROGPOSITE」。除了照片中的黑色外，還有迷彩、深藍
×灰及白色等4色。日本國內則在HEAD PORTER限定販售，
2001年推出。

Released in 1997. Basketball shoe the FOAMPOSITE features a
polyurethane upper surrounded by an unusual mid-sole and
attracted much attention. The CLOGPOSITE is the boldly produced
slip-on version. The model comes in 4 colors, black in the photo,
Digi-Camou (Digital Camouflage), navy and gray, and white. Sold
exclusively at HEADPORTER. Released in 2001.

NIKE
AIR ZOOM SEISMIC

在AIR MAX風潮漸歇的1999年，推出了以革命性設計及高科技為概念的「ALPHA PROJECT」。這款「AIR ZOOM SEISMIC」就是當中的主力鞋款，於2000年上市。「跟同期推出的『KUKINI』、『ZOOM HAVEN』一起，都是我當時很常穿的鞋款。」同年藤原浩也著手推出此鞋的「MONOTONE COLLECTION」版本。

When the AIR MAX boom cooled down in 1999, the ALPHA PROJECT started to pursue improved innovative design and technology. AIR ZOOM SEISMIC was released as one of the project's main models in 2000. "This is just one of the shoes I used to wear. Others include KUKINI and ZOOM HAVEN." In the same year, a version of this model was released as part of the monotone collection designed by Hiroshi Fujiwara.

NIKE
INTERNATIONALIST

1981年推出的慢跑鞋。運用NIKE在1970年代所研發的「WAFFLE SOLE」及「杯狀鞋墊」科技，這款鞋可說是當年慢跑鞋中最完美的代表作。此外，還有藍×黃的配色也十分經典，至今仍有許多NIKE的死忠粉絲把這雙鞋列為最佳代表作。

Running shoe released in 1981. This model features a waffle-sole and a cup insole; an accumulation of NIKE's technology in the 1970's. These shoes were described at the time as "close to perfection" in terms of performance. With beautifully contrasting blue and yellow coloring, many old NIKE fans still pick these out as their favorites.

NIKE and Hiroshi Fujiwara's Relationship

藤原浩與NIKE的關係

從熱情的支持者，變成提供意見的一方。
再到近幾年變成一起合作的夥伴。
藤原浩和NIKE的關係在這十多年間有著巨大的轉變。
能夠讓雙方的價值觀產生共鳴的交會點在哪裡呢？

From passionate fan, to creative contributor with an opinion.
Hiroshi Fujiwara's relationship with Nike has changed dramatically over the past ten to fifteen years.
We take a look at how the two parties with coinciding mutual values came to connect.

從充滿實驗味道的HTM開始、黑白色調系列及fragment design的別注鞋款，藤原浩和NIKE在過往合作過許多不同的計劃。美國Nike, Inc.的董事長兼CEO的Mark Parker，多年來和藤原浩在公私兩方面皆有深厚交情，他又是怎樣看待這個合作的關係呢？

——你是怎麼看待NIKE和藤原浩之間合作的關係？

Mark「我們倆彼此尊敬，已經一起合作超過10年了。浩兄對於設計、造型、文化、科技等各項領域都擁有相當優異的觀察力。這為NIKE提供融合了設計、科技及運動方面的獨到觀點。透過我們的合作，不但能為過

去生產的商品再次注入新生命，也能向外界帶出新研發出的技術。無論是哪種，我們都在很愉快的狀況下合作。」

——藤原先生，你和許多企業都有合作關係，比較之下你覺得跟NIKE的合作有何不同之處？

藤原「因為NIKE是擁有強力後援的大型企業，所以才能進行像HTM這樣具有實驗性的作品，是一份很有價值、很愉快的工作。對於還在研發階段的商品，我也會提出『因為這個技術是如此，如果能夠這樣使用不是挺有趣的嗎？』等等類似這樣的構思。對我個人來說，在商品還沒有正式推出之前，可以比任何人都早見

With experimental project HTM at the top of the list, Hiroshi Fujiwara has contributed to an array of collaborations over the years, including the Monotone Collection and fragment design. We asked President and CEO of Nike, Inc., Mark Parker how he perceives his relationship with Hiroshi Fujiwara, with whom he has come interact with on both a social and professional level.

— **How would you define Nike's collaborations with Hiroshi Fujiwara?**
Mark: Our collaboration goes back around ten years. We have a great deal of mutual respect. Hiroshi has great insights into style and design, popular culture, technology, so his insights combined with Nike and our design, our technology, our connection to sports, creates a new and unique point of view on design. So we have lots of fun sharing ideas and trying to take things that are a very strong design, and make them even better. Taking ideas from the past and bringing them back in a more contemporary form, and taking new ideas and new technologies and introducing them for the first time, through HTM and our collaboration together.

— **Hiroshi, you have a lot of experience working with various enterprises. How would you compare your work with Nike to other companies?**
Hiroshi: Nike is very firmly rooted in sports, but at the same time, they are a major company, so they are able to experiment with creative projects such as HTM. It's highly motivating work. When we develop new products, we're able

到，感覺就好像間諜般，也是很吸引我的地方（笑）。」

——剛剛所提到的**HTM**，由大企業的高層親自參與並主導的實驗計劃，實在非常罕見呢。

Mark「的確NIKE是間大型企業，但我認為同時也必需擁有相對的柔軟性是很重要的，而且有時候甚至要像小型公司一樣，有著快速的應變能力，這也是必要的。」

藤原「我第一次見到Mark時，還是他在當副總裁的時候我問他『NIKE還有什麼想做的事嗎？』他就提出了HTM最初的計劃。這就是HTM這個計劃開始推動的關鍵。」

Mark「我本身就是當設計師出身

的，今後也想繼續擔任創意的角色，不可能因為當上CEO，就把創意丟到一邊。」

藤原「我工作時總是沒有把『企業』當作對象，而是把『人』當成對象，所以對我來說，Mark是一個從事創意的人，然後剛好同時擔任大企業的社長而已。」

——和NIKE合作後，藤原先生對於NIKE的看法有沒有改變？

藤原「沒有改變。」

——有更喜歡NIKE嗎？

藤原「或許有。有時候我會想『其實用不著刻意推出這麼冒險的產品

to think up interesting ways to incorporate (Nike's) technology, which is a lot of fun. Personally I also enjoy being able to see new products before they are released into the market before anyone else. Kind of like a spy. (laughing)

—— Speaking of HTM, isn't it rare for the executive of a major company like Nike to have such an active role in an experimental project such as HTM?

Mark: I think it's important for a big company to be nimble, quick and to be connected like a small company. It allows us to have our own agenda.

Hiroshi: The first time I met Mark he was still vice-president. He asked me if I was interested in working together with Nike on something, and that's when I first started talking to him about the concept of HTM. That was how the project started.

Mark: I'm originally a designer and as a designer, it's important for me to be creative. Because I am CEO (now), it doesn't mean my creativity has gone away.

Hiroshi: I prefer to think of my counterparts as people on a personal level, as opposed to a company. In that sense, I think of Mark as a man with a great sense of creativity, who just happened to become president of a major company.

—— Hiroshi, has your perspective of Nike changed since you started working with the brand?

Hiroshi: No, not at all.

啊」，但是能夠做這樣的實驗，也只有NIKE的實力才可辦到。」

Mark「我總是常常想著要玩新的點子，HTM就是其中一個很好的舞台給我們展現出來。我和藤原浩以及Tinker Hatfield最喜歡的就是嘗試新的設計。」

藤原「因為有了HTM這個名稱，某種程度來說，我們要做什麼都可以。周圍的人也會認為『因為是HTM嘛，也沒有辦法』，甚至因為是HTM所推出的商品，所以大家都抱著高度的好奇。」

——接著請談談即將推出的HTM2？

Mark「這是浩君、Tinker加上我及設計師Mark Smith所組成的新計劃。靈活運用所有NIKE的科技及技術，設計出能夠適合每天穿著、融入日常生活中的鞋款。此鞋款開發與Mark Smith有著極大的關係，今後也預計推出不同的色系，HTM2本身並不是個長期經營的計劃，最多只能算是一個短暫的實驗專案。」

藤原「我認為今後會變成HTM+邀請不同的特別來賓加入設計的方式也說不定，比方說Atsuyo加入的話就會變成HTMA，或是榎本加入的話就變成HTME（笑）。」

——**Mark**對東京的球鞋文化有什麼看法？

Mark「首先是東京的消費者對球鞋實在太熟悉及太有經驗了，眼光十分

—— Have you grown to like Nike more?

Hiroshi: Yeah. At times I think some of their products are a little too adventurous, but having the capacity to be able to experiment is one of Nike's strong points

Mark: We (Hiroshi and I) like to try new ideas. We both love design, and Tinker as well of course, so HTM is a way for us to experiment and test new ideas.

Hiroshi: With the HTM title, we are at liberty to try anything we want. Some perhaps prefer to leave us to our own devices, whereas others are immediately interested in our designs, because they know HTM and appreciate our work.

—— Can you tell us a bit more about HTM2 which is up for release soon?

Mark: Mark Smith, was involved with the design and the development of the shoe. Hiroshi, Mark Smith, Tinker, myself worked together to refine the shoe and bring it to its current state. We started with an original design. It was a unique idea. And something that we felt was very different but relevant for everyday wear. Very simple, easy to slip on and off, very comfortable, using Nike free technology. Performance comes to lifestyle. We'll have new colours, but we are not changing the name (of HTM) permanently to HTM2.

Hiroshi: In the future, new project titles may take the same form of HTM + guest designer. HTMA for Atsuyo for example. Or HTME for Enomoto! (laughing)

銳利，而且他們對於所有的細節都不會放過，還有是不時會有批評的聲音。我們非常理解這個情況，對商品的設計來說，就是要挑戰最終極的水準和品質吧。如同『御宅族』這個很流行的用詞，日本的消費者就是會在特定的事物上追根究底不斷鑽研下去。這點和我自己也有共通之處，我自己本身就是一個集球鞋、玩具、動漫、藝術、科技、設計……的御宅族。東京的球鞋文化與別不同，情況就像點了火般瞬間地爆發起來，和倫敦、洛杉磯、紐約等其他都市相比下真的非常不同。我從1980年代開始，這30年間持續往來東京，每次都受到相當大的刺激而返國。無論在設計層面或是文化層面上也是如此，而且就像和浩君這樣的創意人物

談話後，更能獲得超強的靈感與火花。」

藤原「球鞋本來就是為了運動員而設的，所以NIKE的方向也是一樣，基本上也是考慮到運動員的需求而設計出不同的商品。但是在另一方面，尤其在東京這種和體育界無太大關係的地方，也有顯著的市場存在。比方說，對籃球運動完全沒有興趣的人，也會穿著籃球鞋上街。NIKE的球鞋已經變成人們生活中的一部分了，對於這點我認為十分有趣。」

Mark「我認為最重要的地方在於NIKE的球鞋在真正運動而設計的同時，也務求能為運動員在體育競賽時提高效率。在這個製造過程中，就會產生出生出很多有趣的設計，縱使它原本並非為了時尚的角度出發而設

—— What are your thoughts on sneaker culture in Tokyo?

Mark: Well I think that first of all that the Tokyo consumer is very sophisticated, very discerning, very critical, very high attention to detail. I think it's important that we recognize that almost obsession with quality and detail and we bring that to the design of the products that we create. It sets a very high standard. The word in Japanese "otaku", (meaning someone who has deep focus and obsession), is something I personally relate to. I'm an otaku of sneakers, toys, technology, art, and design, so I can relate. I also like anime, and some of the early animation coming out of Japan, which later turned into a big focus on design. The sneaker culture I think really exploded here in Japan and other major cities like London, Los Angeles, New York and Philadelphia. But I think Tokyo lit a match and it blew up. I've been coming to Tokyo for thirty years, and I've always found it to be incredibly inspiring from a design standpoint, a cultural stand point and when I can connect with talented people, creative people like Hiroshi, it makes it even more exciting and interesting.

Hiroshi: Essentially sneakers are meant for athletes, so Nike utilizes their creativity to design shoes for sports. Despite that, especially in Tokyo, they have established a market for themselves where sport is irrelevant. You'll see kids in the street wearing basketball shoes even though they have absolutely no interest in basketball, yet Nike shoes have become a part of their everyday life.

計，但由於鞋子本身就是一雙貨真價實的運動鞋，所以其設計的特點就會自然呈現出來，而並非只會單純展現出時尚的一面。」

藤原「雖然我知道有某部分的人只會以時尚的眼光來看待NIKE的出品，但實際上NIKE是一直努力創造出追求良好品質的運動鞋品牌，這就是NIKE最棒的精神所在。」

Mark「其實往往在解決功能性問題的同時，球鞋造型的美感就自然會流露出來了。我認為這是一種獨特的設計過程，就如要用哪一種材質作出搭配或是要用什麼方式來組合不同類型的鞋款，我認為只以時尚的角度來思考是不能解決問題的。」

——今後NIKE和藤原浩之間的關係會如何呢？

Mark「今後也會和現在一樣，持續推出好的合作商品。浩君有著深入的觀察力，希望能透過他持續獲得在設計、科技、造形、文化各個層面的接觸，而且接下來除了東京以外，我們希望能夠放眼全世界。對於這些事物上的關心及熱情是我們之間共通的意念，所以彼此的溝通十分好。」

藤原「對我來說，今後希望能做出『看得見設計者的想法』、『能夠引發某種事件』的鞋款。相信能夠回應我的需求的，就只有NIKE，希望日後能和NIKE聯手做出我想做的東西。」

Mark: That's a very important point. Nike design is based on authenticity and performance. It's an obsession to help athletes to perform better. Through that process you create a unique design that's not fashionable for fashion's sake, but because it's real and authentic. I think it has a stronger design presence than just pure fashion, because it's coming from something real.

Hiroshi: There are a lot of people that view Nike products from a fashion perspective, but in fact Nike work very had to create a great product, with little regard to fashion. That's what's interesting about Nike.

Mark: When you solve functional problems, you create a unique aesthetic through the process of solving those problems. What the material looks like, how a pattern is put together, and those are things that could never come out of a pure approach to fashion.

—— What do you expect from Hiroshi, and his relationship with Nike in the future?

Mark: My expectation of Hiroshi is that he continues to do what he does very well. (He has) very deep insights into design and technology, style, youth culture, popular culture. Not just here in Tokyo, but from a global perspective. And we share that passion, so it makes our interaction very easy.

Hiroshi: For me, I want to continue making shoes that make a personal statement, shoes that trigger the next generation. The only brand capable of that is Nike. I want to make those kind of shoes together with Nike in the future.

Mark Parker

美國 NIKE Inc.董事長兼 CEO。 1979年加入NIKE以來，擔任約30年的設計師後，轉任Global Footwear部門的副總裁。 之後擔任 NIKE Brand社長，2006年擔任現職，和藤原浩是從擔任副總裁以來就培養出跨越公私的情誼。

President and CEO of NIKE, inc. Joined Nike in 1979 as one of Nike's first footwear designers back in 1979. For more than 30 years, he's brought innovative concepts and engineering expertise into such vital roles as Vice President of Consumer Product Marketing, Vice President of Global Footwear and Co-President of the Nike Brand. He's led the way for Nike Air and a multitude of industry-breakthroughs in product design. He became President of the NIKE, inc in 2006. He has had a social and professional association with Hiroshi Fujiwara for over ten years.

Hiroshi Fujiwara in Asia

NIKE SPORTSWEAR 2010 S/S in Hong Kong

藤原浩在亞洲球鞋市場
的存在地位

Hiroshi Fujiwara's Presence in Asia's Sneaker Scene

2009年末，藤原浩去了香港，目的是為了參加NIKE SPORTSWEAR 2010年春夏媒體發表會。受到NIKE CHINA邀請的藤原浩，受到香港當地的媒體包圍以外，連中國和台灣的媒體都爭相訪問他。

近年在亞洲圈裡，東京街頭文化有著相當高的指標性地位，當然藤原浩也不會例外，他被當成「東京」、「原宿」的代表人物，在亞洲各地都有相當高的知名度。（題外話，亞洲媒體稱藤原浩為「原宿教父」。）

只要是跟藤原浩有關的服飾，如visvim的服裝及Levi's FENOM丹寧褲，到fragment design推出的NIKE球鞋等等，很多媒體圈的採訪人員都有穿著，在發表會場裡由藤原浩所擔綱的「FOOTSCAPE」及「TENNIS CLASSIC」也引起媒體的高度關注。

Towards the end of 2009, Hiroshi Fujiwara was in Hong Kong to attend the Nike Sportswear 2010 Spring Summer Greater China Media Summit. Upon receiving a request to attend from Nike China, Hiroshi Fujiwara conducted interviews at the event with the local press, as well as media from China and Taiwan.
Tokyo's street culture has grown extremely popular throughout Asia in recent years, and Hiroshi Fujiwara is no exception. In fact Hiroshi Fujiwara is well-known in Asia as an icon of Tokyo and Harajuku. (The press in Asia refer to Fujiwara as the Godfather of Harajuku culture.)
Many members of the press were wearing visvim, Levi's Fenom Jeans fragment design vs Nike shoes and other items closely connected to Hiroshi Fujiwara. Items Hiroshi Fujiwara has worked on recently such as the Footscape and Tennis Classic were drawing a lot of attention from the media at the press conference.

陳冠希解讀
藤原浩的創意

Edison Chen on the Creations of Hiroshi Fujiwara

身為香港創意集團「CLOT」成員之一的陳冠希，本身也是知名球鞋愛好者，也是藤原浩的球鞋大粉絲之一。陳冠希解讀了藤原浩球鞋所擁有的魅力。

「藤原先生是我尊敬的球鞋設計者，因為他所加入的細節低調得來同時能凸顯不同的細節，最厲害是可以保持細緻且簡潔的效果，絕對不會妨礙球鞋本身已經擁有的優點，而且還能把優點進一步提升，這些都是我非常欣賞的地方。」

在藤原浩所擔綱的球鞋裡，他最喜歡的是CONVERSE × fragment design的「ALL STAR」、NIKE × fragment design 的「AIR FORCE 1」、HTM 的「AIR WOVEN BOOTS」這3款。
「藤原先生所設計的全黑『ALL STAR』，非常簡約，和我個人的風格十分配合。無論是隨興或正式的裝扮都能搭配。fragment design的『AIR FORCE 1』是第一次和藤原先生見面時收到的禮物，非常珍貴，現在也是我最喜歡的球鞋之一。HTM的『WOVEN BOOTS』則是讓我掉入球鞋世界的關鍵，在那之前我只喜歡穿『JORDAN』或『DUNK』等款式，遇見這雙『WOVEN BOOTS』後，讓我對球鞋的價值觀有了重大的改變。」

透過藤原浩的手，從東京誕生的球鞋能夠飄洋過海，讓陳冠希這類創意人都為之瘋狂，亞洲各地都出現死忠的追隨者。這份影響力到底能夠蔓延到何種程度，且讓我們拭目以待。

Member of Hong Kong's creative group CLOT and reputable sneaker freak Edison Chen is a devoted fan of the sneakers Fujiwara creates. We asked Chen where exactly the appeal of Fujiwara's sneakers lies.
"Hiroshi is a great shoe designer as he adds subtle but very noticeable highlights to the shoe but still keeps the simplicity. It's like his design compliments the original thought of the shoe, but at the same time doesn't over power the concept of the shoe itself."
Among the shoes Hiroshi Fujiwara's worked on, Chen particularly likes Allstar from Converse x fragment design, Air Force 1 from Nike x fragment design and Air Woven Boots from HTM.
"The Hiroshi Converse's are one of my favourites because they're simple but very sleek. I can wear these to chill on the weekend with friends or I can wear them to most meetings with some khaki's or dress pants."
"The Hiroshi fragment design af1's were a shoe I always wanted. When I met HF they had just come out and I was blessed to get a pair. These are still one of my all time favourite shoes and I remember feeling so lucky to have these when I got them."
"The Air Wovens are my favourites because Air Woven got me into sneakers. I was into sneakers before like Jordans but not like this. I love these."
Shoes designed in Tokyo by Hiroshi Fujiwara have crossed the oceans to capture creators such as Edison, becoming increasingly in demand in all parts of Asia. Just how far will Hiroshi Fujiwara's influence spread? Stay tuned to find out.

CONVERSE
ALL STAR

CONVERSE「ALL STAR」是跨越世代的鞋款，說不定每個人都曾經擁有一雙。從1917年誕生至今，「ALL STAR」的橡膠鞋頭和鞋身側面大大的星星徽章，這個基本的設計從來沒有作出巨幅的改變。換句話說，可以說是沒有必要作出變化，因為最初的設計就已經接近完美了。雖然藤原浩本人年輕時就很愛穿CONVERSE，但由他主導的fragment design卻要在2007年才製作出特別配色的版本。關於他這次的概念，藤原浩本人的說明如下：「和普通CONVERSE不同的地方，在於鞋面、鞋頭、鞋帶孔及鞋跟的徽章圖案設計，全部改用相同的單一顏色搭配。如此徹底的簡化配色，相信是『ALL STAR』至今為止從未實現過的手法。」

A great majority of people across many generations, have at some point slipped on a pair of CONVERSE ALLSTARs. From the rubber toe guard, to the star emblem on the side, there has hardly been any change made to the shoe since the model first appeared in 1917. To put it another way, the shoe was of such a superior design, that there was never any need for changes. Hiroshi among others wore the shoes for many years, and in 2007, he brought out his own unique model of the shoe, under his own brand, FRAGMENT DESIGN. Hiroshi describes the concept of the design in his own words. "The simple difference between the standard ALLSTAR and our design, was that the upper, toe guard, shoelace holes and heel patch were all the same color. I don't think there had ever been such a simple design brought out till that point."

皆為fragment design所設計的款式，鞋跟側面有著跟NIKE合作時所用的fragment design Logo，以刺繡方式呈現。2007年推出，但因為諸多理由，沒有在日本上市。

FRAGMENT DESIGN briefed every detail of this particular version. The same FRAGMENT DESIGN logo that was used in other collaborations which the brand had done with NIKE, is embroidered on the side of the shoe. The shoe was released in 2007, but for various reasons, was never sold in Japan.

Archives | CONVERSE | ALL STAR

CONVERSE
JACK PURCELL

和CONVERSE的定番鞋款「ALL STAR」並駕齊驅的「JACK PURCELL」，原本是羽毛球選手Jack Purcell的聯名鞋款，於1935年面世，因為簡潔的外形和可供上街穿著的舒適感而大獲好評，受到許多球鞋愛好者的喜愛。它擁有超強人氣的理由，是因為它的鞋頭有如微笑記號的線條，及鞋跟處有如鬍鬚般的拼貼三角形圖案，這些具有特色的細節令人感到非常吸引。有著這樣典故的這雙「JACK PURCELL」，也是藤原浩在1980年代時十分愛用的鞋款之一，「相比起『ALL STAR』我更喜歡『JACK PURCELL』，因為它看上去自然又不做作的兩層鞋頭的設計，充滿份量感。」

This shoe alongside the ALLSTAR is the second of CONVERSE's big-time two models. Originally this shoe debuted in 1935 as JACK PURCELL (a badminton player's signature model), but due to its simple design and comfortable fit, the shoe also became popular as an everyday street shoe. The popularity of the shoe has meant that detailing such as the line across the toe guard and the heal patch have come to be known respectively as the "smile" and the "mustache" of the shoe. Drawn to such endearing features, Hiroshi also wore these shoes during the 1980s. "If I had to choose I would say I prefer JACK PURCELL to ALLSTAR. I particularly like the double-layered, chunky toe guard."

皆為藤原浩在1980年穿過的款式，鞋頭的形狀和車線等細節的處理，因為時代的轉變而有著些微的差異。

Hiroshi wore both shoes mostly during the 1980s. The shape of the toe and stitching varies slightly from generation to generation.

Archives

CONVERSE

JACK PURCELL

Gravis
RIVAL

從「BURTON」旗下的鞋子部門於1999年成立的品牌「Gravis」，
以從未曾出現過的「休閒」風格為主要概念，設計出多款嶄新的
鞋款。在品牌成立之時，藤原浩也提供了相當多的意見。「像
Gravis這樣能做出界於球鞋和健行鞋之間鞋款的品牌，在當時是
相當少見的，而且當時市面上的球鞋品牌不像現在這麼多，所
以新品牌的建立也是一件很新鮮的事。」此品牌的當家代表鞋
款就是「RIVAL」，也由於藤原浩曾在雜誌上介紹，此鞋款因而
引起爆炸性的話題。「在Gravis的鞋款中我個人最喜歡的就是
『RIVAL』。」

BURTON launched GRAVIS footwear in 1999. Based on their new and
unique concept of creating comfortable shoes to wear after a long day of
riding, they have continued to bring out a great number of fresh designs.
The launch of this brand prompted a lot of new ideas for Hiroshi. "Shoes
like the ones GRAVIS were bringing out - half sneaker, half trekking shoe
- were unusual for the time. What's more, there were not as many
sneaker brands as there are now, so it was a breath of fresh air to see
the launch of a new brand. The RIVAL is arguably the brand's most
prominent model. Hiroshi helped to introduce the model in magazines,
and subsequently it received explosive acclaim. "For me personally the
RIVAL is my favorite model from the GRAVIS brand."

Gravis和goodenough合作的別注版鞋款，第二版於
2002年推出。

A collaboration model between GRAVIS and
GOODENOUGH. The second edition released in 2002.

以每季更新的節奏推出新款的「RIVAL」，有時會根據不
同時期在細節上作出些微變化。

The RIVAL is updated periodically. Depending on the
release date, some changes in the detailing can be
seen.

Gravis
KONA

「RIVAL」之後的人氣作品，就是這款短筒的「KONA」。在當時
Gravis主要以皮革材質製作球鞋，這款「KONA」卻意外地採用
網狀材質而成為相當獨特的代表鞋款。「KONA」有著透氣性鞋
面及輕量性的兩大特點，這款鞋也是藤原浩所喜愛的作品之一，
據說他個人擁有數雙不同顏色的版本。

The next successful style from GRAVIS after the RIVAL was the KONA.
This shoe with a mesh upper was a rare design from GRAVIS who were
bringing out mostly leather shoes at the time. The breathable and
lightweight properties of the material were the key characteristics of the
design. This model is one of Hiroshi's favorites, which he owns in a
number of colors.

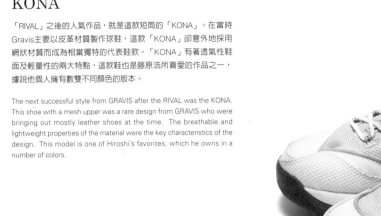

Gravis初期的熱門鞋之一「KONA」。鞋身採用輕量且
高透氣性的網狀材質，因為穿著時十分舒適而獲得很
好的評價。

One of GRAVIS's first big-time models was the KONA.
A light, breathable mesh was used for the upper. The
comfortable fit of the shoe has earned the model
widespread acclaim.

visvim和goodenough合作的鞋款，2001年推出。

A Collaboration model created by VISVIM and GOODENOUGH. Released in 2001.

visvim
TWOMBLY

visvim的代表作「TWOMBLY」是此品牌第一季（2001年）推出的定番鞋款。鞋身正面採用U字形的莫卡辛（Moccasin）縫製法，帶出帆船鞋的外形，再與輕量的球鞋鞋底結合起來。這劃時代的概念，是在visvim品牌成立之初由藤原向中村Hiroki所提出的想法。「TWOMBLY」這個名字的由來是借用當代藝術家Cy Twombly的姓氏。

The TWOMBLY from VISVIM is one of the brand's most popular models, released in the first season during 2001. The shoe features a U-shaped moccasin stitching on a loafer style upper with a light sneaker sole. This groundbreaking concept originated from an idea Hiroshi put to Hiroki Nakamura when the brand was being prepared for launch. The model was named after the modern artist, Cy Twombly.

visvim
HOCKNEY

相對「TWOMBLY」，「HOCKNEY」是一款有如帆船鞋般簡單一套便可穿上的鞋款，是以3個鞋帶孔加上莫卡辛（Moccasin）縫製法的平底帆船鞋款為基礎。至今為止推出過很多不同配色，也和其他品牌有過許多合作。藤原浩本人則喜歡黑色和海軍藍的簡單配色，也是他最常穿的款式。此款鞋的命名由來則是借用了當代藝術家David Hockney的姓氏。

The HOCKNEY is a laced version of the slip-on shoe the TWOMBLY: It's a moccasin deck shoe that has three eyelets. The shoe has been released in a number of color variations and has also been the subject of collaborations with other brands, but Hiroshi preferred to wear this model in simple black and navy. The name for this model originates from the modern artist, David Hockney.

皆為HOCKNEY初期的鞋款，正確推出年代不詳，約為2005~2006左右。

Both early models of HOCKNEY. The official release date is unknown, but is thought to be sometime in 2005-06.

visvim
FBT

以「能穿出街頭感覺的印第安莫卡辛鞋」的概念得出的設計，「FBT」以莫卡辛鞋面加上球鞋的鞋底結合而成，是最能代表visvim的代表鞋款之一。自2001年第一季推出以來，在功能性及鞋款設計上不斷追求改良，持續推出更新的鞋款及不同的別注版本，達到今天的成果。此外，當初這個概念是由藤原浩所提出的，而取名「FBT」則是來自音樂團體FUN BOY THREE的名稱縮寫。

Based on the concept, "Indian moccasins for street wear", the FBT comprises a moccasin upper mounted on a leather sole and is VISVIM's magnum opus. The shoe has continued to evolve since it was first released in 2001, particularly in the technology of the design. The shoe was conceptualized by Hiroshi and named after the band FUN BOY THREE.

歷代的FBT裡，藤原浩最喜愛的就是這雙以皮革做鞋面的款式，他平日穿著時更會把流蘇的部分拆掉。

Hiroshi's no.1 choice from previous FBT lines, is this smooth leather type. He removed the fringe when he wore them.

帶有流蘇的這款是和UNDERCOVER合作的別注版本。

This model with fringe is a collaborative version with UNDERCOVER.

「FBT」的初期鞋款。藤原浩加上goro's的南美風格披風（Poncho），變成獨一無二的自我改裝鞋款。

First model of FBT. Hiroshi added a poncho from Goro's to come up with this unique customization.

visvim
SERRA

以戶外爬山靴形態設計的「SERRA」，也是visvim的代表作之一。外觀雖然看起來略顯沈重，但實際上卻採用了輕量化的材質製造，是一雙非常適合在都市內行走穿著的鞋款。藤原浩本人很喜歡這雙鞋，也曾一度經常穿著，過去曾經和fragment design合作推出限定鞋款。「這雙鞋和丹寧褲十分搭配，是我冬天的必備鞋款。」

This twist on traditional trekking boots named the SERRA, is another popular model from VISVIM. The shoe appears to be heavy duty, but a lightweight material has been used for the sole, making the shoe suitable for street wear. Hiroshi wore this shoe in the past and also released a limited edition collaboration model of the shoe under his label FRAGMENT DESIGN. "These shoes go well with denim. For me they were a must for winter."

高橋盾和藤原浩聯手的AFFA和visvim共同合作的鞋款，大膽地採用了偏大的格紋圖案。

A collaboration model between AFFA and VISVIM, co-designed by Jun Takahashi and Hiroshi, featuring a bold check pattern.

visvim和fragment design合作鞋款，鞋身採用白色皮
革，鞋舌部分則壓上fragment design的閃電Logo。

A collaboration model created by VISVIM and FRAGMENT
DESIGN. Features the FRAGMENT DESIGN trademark
on the tongue.

Dialogue
Hiroshi Fujiwara × Hiroki Nakamura (visvim)
對談 藤原浩×中村 Hiroki（visvim）

以鞋出發從中發揮獨到的藝術品味，visvim在日本國內及海外都擁有高度評價。
擔任總監的中村Hiroki和藤原浩份屬多年好友。
在藤原浩的眼裡，現在的visvim究竟是什麼模樣的呢？
從各自創作的觀點到彼此之間的友誼，2人之間的對談馬上展開。

visvim has established an artistic collection mainly of shoes,
which has been well received both within Japan and overseas.
Director Hiroki Nakamura and Hiroshi Fujiwara are old friends.
How does Hiroshi Fujiwara perceive visvim today?
We discuss with Hiroshi and Nakamura their thoughts on design and creation,
and hear more about their personal relationship.

——藤原先生，可否用你的角度去談談今天的visvim？

藤原「不只日本國內市場，visvim整個品牌都有著國際大都會的感覺，氣勢直線上升當中。感覺上在海外比在日本國內還要受歡迎呢。」

——visvim從2009年開始也開始至巴黎舉行展示會，在歐美得到什麼樣的評價？

中村「我想因為visvim是歐洲和美國所沒有的品牌類型，所以才因此獲得很好的評價吧。現在和以前不同，所謂的『大都會市場』全世界走到哪兒都有，visvim只是有種變成『共通語言』而被理解接受的感覺……」

——中村你自己在開展visvim時，視野本來就不只日本，也已經包括了海外是嗎?

中村「是的，成立visvim的重點就是要製造出好穿實用的鞋子，並沒有特別針對日本國內的需求而設計。」

——雖然製造出來的效果也出現了「日本味」、「東京味」的風格，但這應該不是你當初所設定的方向吧？

中村「『帶有東京味』這樣的因子，應該是自然流露出來的，並不是我最初設定的目標。」

——本書介紹了藤原浩過去所穿過及參與製作過的鞋款，中村你又如何看待這些內容呢? 你有沒有什麼以往曾穿過的好鞋呢？

中村「年輕的時候因為有玩滑板，所以藤原浩在雜誌介紹過的鞋款我都會穿，就如VANS和NIKE等等。但書內不管哪雙球鞋，都很有『藤原風格』啊。」

——到底什麼是「藤原風格」呢？可以具體地說明嗎？基本上藤原所穿的或創造的鞋款都以簡潔為主，你認為當中有什麼元素貫穿整個風格呢？

中村「就是那個『不知道是什麼』才正是『藤原風格』啊！用絕妙的手法，著眼在大家都尚未察覺的小細節上，這個功力真是沒人可以學得來。藤原浩在找出『似有若無』的東西上真的很強，只要變換角度竟然就能產生這麼大的不同及變化。看著藤原浩過去創作的東西，我也不自覺地受到影響。但就算我用相同的手法，也絕對無法與他匹敵。」

——大家都知道藤原和NIKE的關係深厚，而中村你之前也曾在BURTON內部工作。想必你對在大公司內要挑戰試做新商品的困難有深刻體會; 關於這點你有什麼想法？尤其是像藤原的狀況，他是直接和社長一起進行新計劃的。

中村「如果不是用這種方法，就不可能會有像HTM這麼獨特的商品系列出現。和具有主導權的人直接工作，對於把藤原的想法具體化實現，是一件非常有利的事。當然，NIKE和BURTON這兩家公司的規模及公司文化根本不一樣就是了。」
藤原「大企業裡如果滲入政治因素，通常不能推出有趣的東西。」

——Hiroshi, can you tell us a bit about visvim today, from your own point of view?
Hiroshi: "Without limiting themselves to just the Japan market, I feel they have matured quite rapidly to become a cosmopolitan brand. In fact their reputation is more solid overseas than it is in Japan."
——visvim has been holding exhibitions in Paris since 2009. What has been the response in the West?
Nakamura: "visvim is a brand that brings something very new to Europe and America, so in that sense we have been very well-received. Unlike the past, "cosmopolitan markets" can be found in all parts of the world. All people need to understand each other now is a common denominator."
——Did you yourself intend to take visvim overseas when you first started out?
Nakamura: "We were focusing our efforts on creating comfortable and wearable shoes, so we weren't really thinking about adapting our shoes to suit the markets of countries overseas."
——So even though the shoes may have been received as "very Japanese" or "Tokyo style", that wasn't what you were aiming for as such?
Nakamura: "Naturally our shoes may appear "Tokyo style", but it's not something we purposefully tried to bring out."
——In this publication, we are introducing some of the shoes Hiroshi wore and designed in the past. What do you think of the collection? Did you wear any of the shoes yourself?
Nakamura: "When I was young, back when I was skateboarding, I often wore the shoes Hiroshi introduced in magazines, such as Vans, Nike… All the sneakers had a certain hint of Hiroshi to them."
——How would you define "that hint of Hiroshi"? Generally speaking the shoes Hiroshi has worn and created are fairly simple in design. Do you think there is an underlying something that these shoes have in common?
Nakamura: "It's exactly that indefinable something about a pair of shoes that gives it "that hint of Hiroshi". Nobody could imitate Hiroshi's talent for noticing the subtle qualities of a product before anyone else. Hiroshi is really good at finding the underlying quality of things. He can teach you a whole new way to look at something, just by changing the angle you look at it from. I think to some extent I have been naturally influenced by Hiroshi over the years, but I don't think I could ever live up to him using the same methods he has used."
——Hiroshi has a strong relationship with Nike, and has been working on a project directly with the Nike CEO. You also worked for Burton in the past, so I'm sure you are aware of the difficulty to challenge yourself to do something new within a company. Can you share with us your thoughts on this?
Nakamura: "It's difficult for unique products (such as HTM) to take off without receiving such support. Being able to work directly with a figure that has such enormous decision making power, is definitely a key factor that will help Hiroshi to put his ideas into action. Having said that, Nike and Burton are very different companies both in size and culture."
Hiroshi: "I agree it's difficult to come up with new interesting ideas if you are all caught up in the politics of a large company."
——We have heard that Hiroshi contributed some design ideas when visvim was planning their first lineup. Does Hiroshi still have a working relationship with visvim?

——初期visvim的商品裡經常有藤原浩的點子，之後又如何呢？現在藤原浩和visvim的關係是怎樣？

藤原「現在沒有關係了，我偶爾去看一下他們的展示會，就這樣。」
中村「可是最近你都不來了……」

——過去曾經有過合作的鞋款推出吧？

中村「藤原會給我意見，這雙鞋的這裡應該要這樣比較好，譬如說這裡的刺繡要用像牙白色等等。」
藤原「其實我們一直有想要聯手做出些什麼出來，但是實在太忙了，所以一直無緣實現。」
中村「那要從零開始思考，還得做很多功課。」

——具體來說會是什麼樣的商品呢？

中村「還不能講。」
藤原「在鞋跟部位裝上輪子，走到哪都能滑行的鞋子……」
中村「如果真的出這個商品，那衝擊性可大了（笑）。」

——（笑）還有，藤原浩的半生傳記《山丘上的PUNK》裡，中村有寫道

「藤原浩即使提供意見，也不收酬勞」這件事。

中村「是的，真的是這樣。」
藤原「關於錢，有能力支付不是更快樂嗎？所以對我來說，是一種『不收錢的私心』。」

——不收錢的私心……嗎？

藤原「我至今為止去過的活動，即使不用入場費，我還是照付。年輕時我因為以來實身份參加，所以都是免費的，這樣當然會很開心，但現在我反而會因為這樣而特別去付費。」
中村「嗯～這還真是心機很深（笑）。」

——真的。（笑）

「不過，中村你為何想做鞋呢？」

中村「我在BURTON工作時已不斷思考想做出『有深度的作品』。深度指的是方便、容易穿著，比服裝更接近『工具』的東西，那麼最能表現這個想法的就是『鞋子』了。」
藤原「就我所知，在日本以獨立方式造鞋的品牌，visvim是第一個。」
中村「在海外有很多，在日本好像真的沒有。」

Hiroshi: "No, not now. Sometimes I will go to the odd exhibition but…"
Nakamura: "Recently he hasn't been coming to many…"
——Is it true that the two of you have collaborated on a product before?
Nakamura: "Hiroshi has given us some suggestions on how we might adapt certain models. For example, changing the colour of the stitching of a shoe to ivory or…"
Hiroshi: "The truth is there are a few ideas that we have began to work on, but I guess because he's busy, he hasn't gotten round to putting any of our ideas into action."
Nakamura: "Well, there's a lot to think about, like how we are going to manufacture a product, so it takes time."
——What ideas have you come up with?
Nakamura: "We can't talk about that just yet."
Hiroshi: "The next visvim shoes will have rollers on the heals, so you can can skid along the streets…"
Nakamura: "Shoes like that would certainly cause a stir! (laugh)"
——(laugh) Nakamura-san, in another publication about Hiroshi, "Tiny Punk on the hills", you commented that "Hiroshi provides us with a lot of ideas, but he never accepts financial compensation for them."
Nakamura: "Yes, it's the truth."
Hiroshi: "I prefer to be on the paying end than the receiving end… I have certain ulterior motives that make me prefer not to receive money."
——Ulterior motives that make you prefer not to receive money?!?!
Hiroshi: "When I go to concerts and events, even if they tell me there is no charge, I try to pay. When I was younger, I felt lucky to be let in for free as a guest, but nowadays I like to pay."
Nakamura: "Hmmm… sounds highly strategical… (laugh)"
——Indeed (laugh)
Hiroshi: "Anyway, Piro (Hiroshi's nick name for Nakamura). Why did you decide you wanted to become a shoe maker?"
Nakamura: "Since I was at Burton, I have always thought I want to make products that had purpose. I'm talking about ease of use, easy to wear, a product that was closer to equipment than simple clothing. That's why I chose shoes."
Hiroshi: "As far as I know, visvim was the first sneaker brand to start off as a one-man operation."
Nakamura: "There were some cases overseas, but I think I was perhaps the first in Japan."
Hiroshi: "Leather shoes are perhaps a different story, but I think you were the first to start off as an individual sneaker maker. At the time to tell you the truth, I had my doubts as to whether you could do it. Compared to standard clothing, there is more

藤原「皮鞋製造可能還有，但以個人品牌開始製造球鞋就真的沒有了。老實說，我也曾在心裡懷疑，這是否可行。和服裝不同，鞋子畢竟有製造數量的壓力。在某種程度上來說，製鞋是必需以長時間持續製造相同的商品，而且也必需跟得上時代流行的腳步。然後每個人的腳形都不同，好穿的鞋也不是人人都會覺得好穿。」

——visvim接下來要推出顧客訂製服務嗎？實際量度顧客的腳圍，再依各人需求一雙一雙地製作出來。

中村「是的。雖然不能為每一位客人量身訂作，但我們會以直營店的顧客為對象開始提供這項服務。」

藤原「說到這裡，我記得以前中村你老嚷著『好想開壽司屋』對吧？壽司屋就跟你理想中的Person to Person的形態很接近了。」

中村「直接和客人接觸，思考著現在他想要吃些什麼，然後立刻在當場具體化地提供他想要的東西。我想這就是服務最終極的目標。然後我想，不知道製鞋是否也能做到相同的服務，才會有這次訂製服務的構想。很久前就有這個想法，不過各方面的條件到現在才完全整合起來。」

藤原「如果NIKE或NIKEiD來做顧客訂製服務的話，又會是完全不同的模式了。那中村你在製作新鞋時，腦袋裡會先有完成圖的形象嗎？」

中村「有啊。創作時最先的就是腦海裡完成的圖像。」

藤原「我也是。不管是什麼，我一定會先在腦袋裡有個完成的形象。前陣子我和朋友聊到這件事，朋友否定我的說法，說也有人是一邊創作一邊思考的。以前我和作曲家筒美京平曾經聊過，他說像他在替櫻田淳子或岩崎宏美等人寫歌時，當歌詞完成後，他的腦海裡就會浮現她們唱這首歌的樣子。當時我聽到後，相當地驚訝並贊同，但後來仔細想想，我們在開始創作前腦袋裡也早已有完成圖了。」

中村「當然，在創作的過程中還是會有新的發現及構想，這時又會出現更新的完成圖。總之不可能在腦袋裡沒有任何想法的情況下開始創作。」

藤原「是啊，不會發生『不想那麼多，先來試試看吧』這種情況。為什麼我會想起這件事，是因為我以前也曾在部落格裡寫過，我會把『煙囪』跟『隧道』搞混。譬如在高速公路上時，我有時會講錯說『等一下走那個煙囪…』。我一直在想為什麼我會弄混，我想是因為這二樣東西在我腦子的形象都是又長又黑的吧。」

中村「這是人類大腦有趣的研究呢。」

藤原「說回來，visvim創立多久？」

to think about on the manufacturing side. Such restrictions compel you to continue to make the same product for a longer period of time, yet you also have to keep up with the trends of the market. What's more, foot shape varies from person to person, which also makes the design process more complex. Even if a shoe is deemed a good fit, it doesn't mean the shoe will fit anyone."

——I heard visvim will start custom order sneakers. Taking measurements of the customer's foot and making a shoe to fit.

Nakamura: "Yes. It's not a service thatwe will be able to provide to all customers, but we are planning to start the service at our directly-managed stores in the near future."

Hiroshi: "That reminds me, you alwayssaid you wanted to become the owner of a sushi restaurant, right? You said that the sushi chef represented your ideal of providing a form of one-on-one entertainment."

Nakamura: "Sushi chefs meet face to face with their customers to work out what they want to eat, and provide exactly the sushi they want all in one sitting. It's the mother of all services. I wondered if we couldn't do the same thing with shoes. And that's how I came up the idea for this custom order service. It's been a concept I have been working on for a while now, and we finally have the resources and setup to make it a reality."

Hiroshi: "Nike have their own custom service, NikeiD, but this is a totally different approach. When you start making a new pair of shoes, do you have a clear picture of the end product in your head?"

Nakamura: "Yeah, when I make something, I always have an image of it first."

Hiroshi: "I'm the same. In anything I do, an image always appears in my head. I was talking about this with some friends recently. Some of them said that they gradually build on ideas as they create. One day I was speaking with song writer, Kyohei Tsutsumi, and he said that for example when he writes songs for artists like Junko Sakurada or Hiromi Iwasaki, once he has written the song and lyrics, he has an idea of how it will sound when they sing his songs, before they start recording. At the time I thought that was amazing, but in fact we're the same because we too have a picture in our heads before we go into the building phase."

Nakamura: "Of course, sometimes new ideas surface even after we have started making the shoe, and then our picture of the end product changes a little. In other words, we will never begin making a shoe before we have an idea of how we want the end result to look."

Hiroshi: "That's right. The reason I remembered that conversation just now is, I often mistake the words "chimney" and "tunnel". For example, I could be driving down the highway, and I'll say by accident, "Take that chimney…". Probably because in my head, they are both long and dark objects."

中村「10年了。」
藤原「我從之前就一直跟你說『也該稍微休息一下了吧』。」

——《山丘上的PUNK》裡，藤原有寫到中村是你朋友中最努力工作的人，是嗎？

藤原「事實上他到底是不是真的在工作就不得而知了……」
中村「在某程度上，我也是一直在玩樂（笑），工作就是興趣嘛。」
藤原「最近總是在全世界各地飛來飛去，很難有機會放假玩樂。」
中村「藤原老是說我開始成立visvim後便因為太忙沒有時間去玩了。」
藤原「你在BURTON工作時我們常常一起出去玩啊，很有趣呢。」

——中村你是一天24小時想著工作嗎？還是會清楚分開ON跟OFF呢？

中村「對我來說，工作和玩樂是一樣的，沒有去分ON或OFF。」
藤原「要分ON或OFF太困難了啦。我就是辦不到，究竟大家是怎麼可以做到的呢？在身上裝個按鈕就可以隨時切換嗎？」
中村「我和藤原雖然在創作上有共通之處，但對於時間的感覺卻完全不同，藤原的速度感比我快，這是因為我本身是品牌製作人的身份，反而會慢慢地工作，像操控著一艘大船一樣，無法快速轉向。」
藤原「我反而是那種任意說出『可以這麼做嗎？』、『如果那樣做不是更有趣嗎？』的人，但從我丟出提案後

Nakamura: "It's funny how the human brain works, isn't it."
Hiroshi: "Anyway, how long has it been since you started visvim?"
Nakamura: "10 years"
Hiroshi: "In the ten years since he started visvim, I've been telling him continuously to take more time off work but..."

——It was also written in "Tiny Punk on the hills", that among the people around Hiroshi, you are the most hard-working.
Hiroshi: "Well, I don't know for sure he's actually working but..."
Nakamura: "In a sense, I'm always playing. (laugh) For me, work is my hobby."
Hiroshi: "He's been gallivanting all over the world, so recently he doesn't have much time to hang out with me."
Nakamura: "When I first started visvim,I remember you saying to me that I would not be able to make time for you because I would get so busy."
Hiroshi: "When you were at Burton, we hung out together a lot and it was fun, right?"
——Are you the type that constantly has work in the back of your mind, or are you able to switch off sometimes?
Nakamura: "For me work and play is all the same thing. I don't really share the concept of on and off."
Hiroshi: "Separating on from off is difficult, isn't it. I'm not good at it either. I wonder how everyone does it."
Nakamura: "Hiroshi and myself are the same in that we both work in design, but our sense of time is totally different. I think Hiroshi works a lot faster than I do. I'm a manufacturer,

Hiroki Nakamura
中村世紀
CUBISM代表，同時是時裝品牌visvim的主理人。中村世紀多年來所做出富藝術感的鞋款，贏得日本國內及海外的一致讚賞及擁戴。www.visvim.tv

Representative of CUBISM. Director of fashion brand visvim. Nakamura's artistic collection of footwear has won him widespread acclaim, both in Japan and overseas. www.visvim.tv

就要開始漫長的航海歷程吧。」

中村「想要開創一個全新的東西，只有2、3個月是辦不到的。一定得用3、4年這種中長期的眼光來思考整體架構才行。」

——visvim今後的展望呢？有預計會推出10周年特別計劃嗎？

中村「目前沒有任何想法。有在考慮是不是要做紀念冊。」

——提到書本，visvim每季推出的型錄都相當有看頭啊。

藤原「以前還曾經有過在KINKO'S影印出刊的時期⋯⋯」

中村「最初是這麼做的沒錯，總共才做20本，一本一本細細地做。」

藤原「就是因為有過那樣的時期，才能有現在呢。」

——最後，請跟我們分享今後visvim的中長期遠景。

中村「基本的方向維持不變，希望做出能反映時代氛圍的商品，品質上做得更好。」

——藤原呢？你對今後的visvim有什麼期待？

藤原「visvim在其它國家都有相當高的評價，我自己也感到自豪呢（笑）。希望visvim未來能夠維持這樣的風格，繼續不斷努力。」

so we work slowly like a captain steering a huge ship. We simply can't turn quickly."

Hiroshi: "My job is to make suggestions like "Can you do this?" or "Wouldn't it be interesting if we could do this?" without thinking too much about what might be involved, or the voyage these guys have to take to make it a reality."

Nakamura: "When I decide I want to try something new, I won't see results in a matter of months. New ideas have to be made into long-term plans that will last for three to four years."

——Are you planning anything for the ten year anniversary of the brand?

Nakamura: "No plans as such at the moment. I have considered bringing out a book, but nothings decided yet."

——visvim bring out quite a magnificentcatalogue every season, isn't that right

Hiroshi: "There was a time they were making them at Kinko's…"

Nakamura: "That's how we did it in the beginning, yes."

Hiroshi: "It's such times that brought you to what you have today."

——Lastly, can you share with us your vision of visvim in the future?

Nakamura: "We won't be changing the way we do things in any big way. I hope our products will continue to reflect the mood of the times, and we will continue to improve the quality of our products."

——Hiroshi, what do you expect from visvim in the future?

Hiroshi: "visvim is very well received wherever it's taken, so I feel very proud of the work they have done. I simply hope they will continue to build on the success they have earnt themselves so far."

new balance
576 & 996

「我認為new balance本身是一個很好的品牌，我自己也很常穿著。」回顧藤原浩穿著new balance的歷史，他偏愛兩個特定的型號，就是「576」和「996」。這兩款都是1988年誕生以來持續熱賣的款式，如今仍舊是常賣款，隨時都能入手的定番型號。藤原本人基本上愛穿的鞋款都是從1990年代開始就穿到現在，鞋身早已陳舊卻仍然愛用，他說：「我雖然喜歡它的穿起來的感覺和外形，但我對new balance這個品牌完全不熟悉。576和996我也不知道這些號碼的由來。」不管產品的背景或歷史是怎麼樣，相信「好東西就是好」是再簡單不過的道理。

"NEW BALANCE is a good brand. I used to wear them a lot." Looking back on Hiroshi's past, you can witness a clear partiality for these two product numbers. Both are popular long sellers since their debut in 1988, and remain standard models, which may still be found in stores today. Hiroshi wore the shoes predominantly in the 1990s until they fell apart, but he claims, "I liked the comfort and the appearance of the shoes, but in fact I don't know much about NEW BALANCE. I don't even know where the numbers 576 and 996 came from." Could this be Hiroshi saying regardless of the background and history of a shoe, a good shoe is a good shoe?

576

在576之中最有人氣的就是棕色系列。深棕色叫
「CORDOVAN」，偏淡的棕色叫「CHOCOLATE」，皆
為Made in USA的出品。

Of the 576 line, the model produced in shades of brown has
been particularly popular. Dark brown is referred to as
"cordovan". Light brown referred to as "chocolate". All
are made in the USA.

996

和576比起來略顯細瘦的外形是其特徵。材質上以網狀
為主，另外也有全皮革的鞋款，但藤原浩本人較喜歡
的是網狀材質的款式，皆為Made in USAB的出品。

Compared to the 576, 996 is slightly thinner in form.
Most shoes in the line are made from mesh, but a few
were made in leather. Hiroshi mostly wore the mesh
version. All shoes from this line were also made in the
USA.

PUMA
CLYDE

1973年登場的籃球鞋「CLYDE」是1968年誕生的PUMA名作「SUEDE」進階版本。其命名由來是1960年代到1970年活躍於NBA的選手Walter "Clyde" Frazier的暱稱「CLYDE」。去除了過多無謂的裝飾，展現簡單純樸的美感，誕生近40年至今仍然相當流行。至今為止，「CLYDE」在這幾十年來推出過許許多多不同材質及顏色搭配的鞋款，藤原浩最喜歡也最常穿的是簡單配色的麂皮材質款式。「這雙鞋的優點在於它和麂皮材質產生相輔相成的協調性。獨特的褪色美感，能讓人從雙腳開始思考全身的服裝搭配。」

The CLYDE appeared first in 1973, as an updated version of the famous PUMA model, the SUEDE. The model was named after the NBA star Walter "Clyde" Frazier, who flourished in the 1960s and 1970s. The beautiful form of this model, free from excessive decoration, has been around for almost forty years now and still shows no signs of fading out. Over the years, the model has been the subject of many color/material variations, but Hiroshi likes the simple coloring of this suede model. "What makes these shoes good, is the suede material used. I also like the unique coloring. These shoes make me want to coordinate my whole outfit from my feet up."

CLYDE雖然多年來擁有無數的配色，但藤原浩卻喜歡穿著側面PUMA標誌和鞋帶同色系的鞋款。

The CLYDE has been brought out in a wide range of colours, but Hiroshi liked to wear this yellow lined and laced version.

adidas
STAN SMITH

貴為歷史名作且在球鞋歷史上佔有重要一席位的adidas「STAN SMITH」，最初在1965年時是以「HAILLET」這個名字上市，之後因為受到網球選手Stan Smith熱愛並且經常穿著，因而改名沿用至今。如此一雙擁有歷史背景的網球鞋，從誕生至今當然推出過許多不同的顏色版本，但藤原浩卻篤定地說：「『STAN SMITH』只有白色最好看」。事實上，藤原本人在1980年代也愛穿白色的「STAN SMITH」。「對我來說，『STAN SMITH』是球鞋界的優等生，簡單的外形看上去彷彿沒什麼大不了，但實際上卻是一雙完成度相當高的經典設計傑作。」

The STAN SMITH, a historical masterpiece from the sneaker history was first launched in 1965 under the name "Haillet". The model was renamed after the famous tennis player Stan Smith at a later date. The shoes have been brought out in a range of different colors over the years, but Hiroshi claims, "White is the only true color for STAN SMITH.". Hiroshi himself often wore the shoes in the 1980s, needless to say in white. "For me STAN SMITH is the "nice guy" shoe from the sneaker world. The shoes may appear simple in form, but in fact the finish on these shoes is extremely high-quality."

酷似「STAN SMITH」的魔鬼氈搭帶版本，但這雙卻是名為「MASTER」的款式，鞋底和「SUPER STAR」採用相同的人字形坑紋。

This shoe exhibits a remarkable resemblance to a velcro version of the STAN SMITH, but in fact this is a separate model called MASTER. A herringbone pattern (similar to the SUPERSTAR) has been used for the outer sole.

以白色為鞋身主色，並在小細節處搭配綠色，帶有濃厚的網球鞋風格。鞋舌處印有Stan Smith的義名圖標。

Typical tennis shoe coloring of a white base, detailed in green. An illustration of Stan Smith has been printed on the tongue.

藤原浩自行改裝版本，在鞋舌上加上他個人畫像及
「HIROSHI」文字的非賣版本。

Hiroshi's own customization, with "HIROSHI" embroidered
on the tongue. Not for sale.

adidas
SUPER STAR

1969年誕生的「SUPER STAR」，其最大的特徵就是橡膠鞋頭的設計。這個具有個性的鞋頭不單只有裝飾的功能，還具有保護前足部和延長鞋子穿著壽命等功能。進入1980年代後，「SUPER STAR」因為Run DMC的愛用而進一步與嘻哈界扯上關係，跨越了單純的籃球鞋領域而成為時尚的元素。藤原浩也曾是愛用者之一，有時會以當時流行的元素穿著，比方說除下鞋帶的穿法等，有時則會充當成滑板鞋，直到鞋面完全磨損為止。

The SUPER STAR was released in 1969. The most prominent feature of its design is the rubber toe cap. Otherwise known as the "shell toe", this element of the shoe is not simply decorative, but has the function of guarding the foot, and increases the durability of the shoe. In the 1980s, the SUPER STAR fused with hip-hop culture when Run DMC started wearing them. This triggered the shoe to evolve from a simple basketball shoe to an icon of street fashion. Hiroshi was also a devoted fan of the shoe, sometimes wearing them without laces (a popular trend of the time) for fashion. He also wore the shoes for skate boarding till the upper was completely worn out.

同樣都是「SUPER STAR」，也會因為生產的年代、製造的國家不同而有著材質、鞋形及細節部分的差異。在狂熱份子的世界裡，1980年代法國製的「金鞋舌」是評價最高的版本。（圖片中的白X綠色即為該款）

The silhouette and detail of the SUPER STAR varies largely depending on when they were released and in which country they were produced. A version of the shoes made in France, know in Japan as "gold tongue" is particularly highly regarded among sneaker freaks. (white and green model photographed here)

藤原浩在1980年代愛用的黑×白「SUPER STAR」，
是歷代鞋款中品質最高的法國製版本。

SUPER STARs in black and white, which Hiroshi wore in the
1980s. Throughout the history of the model, this model
made in France is thought to be highest in quality.

adidas
CAMPUS

1983年誕生的「CAMPUS」，原本是為訓練用而設計的運動鞋，但是和前文提到的經典「STAN SMITH」及「SUPER STAR」一樣，最後變成時尚的代表鞋款而長年受到各界的支持。這雙鞋款也和嘻哈音樂文化有著深厚的淵源，推出後馬上受到BEASTIE BOYS（野獸男孩）的愛戴而聞名。藤原浩在當年也相當喜歡穿著此鞋，「這樣簡單的鞋款，只要在材質和細節方面稍微作出一點改變，整體的印象就會馬上大大不同。『CAMPUS』在過去多年來曾經數度推出不同的復刻版本，但在我心中『CAMPUS』就是麂皮的材質最為對味。我喜歡它獨特配色的感覺，所以收藏了多雙同款但不同顏色的版本。」

The CAMPUS, first launched in 1983, were originally developed for training purposes, but just like its predecessors the STAN SMITH and SUPERSTAR, the model has also grown to be loved as a fashion item. The shoes have been strongly connected to hip hop culture, famously worn by The Beastie Boys soon after the shoes debuted. Hiroshi was also wearing the shoes at the time. "The overall impression of such simply designed shoes, largely changes when small alterations to the material and detailing are applied. The CAMPUS has been reproduced a number of times over the years, but for me this suede version of the shoe is the best. I particularly like the unusual coloring of this version, but I have it in a number of different colors."

1980年代的「CAMPUS」，在2009年曾以「CAMPUS 80's」的鞋名以同樣的材質及鞋形再次推出復刻。

CAMPUS shoes from the 1980s. Replicating the original texture and silhouette of the shoe, "CAMPUS 80s" is a reproduction of the original model, released in 2009.

Dialogue
Hiroshi Fujiwara × Kazuki Kuraishi (adidas)
對談 藤原浩×倉石一樹（**adidas**）

他們共同擁有fragment design這個「Label」，
各自與「NIKE」及「adidas」這對競爭對手有著深厚關係。
藤原浩和倉石一樹，象徵著東京近年具有無窮創造力及開放思想的兩大人物，
一起敞開心胸暢談關於「現在對於球鞋的感受與想法」。

Whilst working as joint partners on fragment design, Hiroshi Fujiwara and Kazuki Kuraishi
also have another strong working relationships with rival companies: Hiroshi with Nike, and Kazuki with Adidas.
The two ambassadors of open-minded creative modern Tokyo, talk casually about their thoughts on sneakers today.

——由倉石先生努力提攜的企劃「adidas Originals by Originals」（以下簡稱ObyO），始於2009年春夏，來到2010年春夏的本季已是第三季了，您有什麼感想？有沒有在創作商品的過程中遇到困難之處呢？

倉石「服飾方面，設計及創意的發揮都比較自由，但相比之下，球鞋則比較侷限。我認為這並不是adidas的關係，因為要開發新的鞋底本來就會對設計做成困難，這是事實。」

——是因為成本的關係嗎？

倉石「是的。重新開一個金屬模型需要不少成本，所以我的做法是以符合現有鞋底的設計為中心。」

——藤原先生，你為NIKE創作球鞋時也遇到相同的問題嗎？

藤原「不完全是這樣。有時候真的只會做鞋底以上的設計，但是也有從鞋底開始重頭設計的時候，不過這種情況花的時間會變得非常多了。」
倉石「我沒有試過從鞋底做起，所以沒有感覺到要花那麼多時間！」
藤原「雖然如此，不過從計畫直到上市發售，整個過程就要不少時間。」
倉石「是啊，要1年以上。」
藤原「可以的話很想快速完成！」

——從藤原先生的觀點來看，倉石先生創作的ObyO反映出什麼信息？

藤原「雖然adidas以運動品牌起家，但是和NIKE等其他運動品牌相比，潮流感是比較強烈的。所以在我來看，一樹君所做的事情我十分理解，應該也可以很順利地進行吧，這方面你覺得是嗎？」

倉石「嗯，雖然進行時沒有特別的困難，但是如果說完全順利，也並不如此……」
藤原「我的感覺是adidas的高層對於一樹君所做的事情，能理解的地方及支持是滿多的。」

——倉石先生負責的ObyO，除球鞋以外，也會設計服飾商品，但是藤原先生是以球鞋為主吧。

藤原「是的。關於服飾方面，我只會偶爾提供一些想法。」

——藤原先生不想認真考慮和NIKE一起設計服飾嗎？

藤原「如果站於現在NIKE的立場來思考的話，有困難啊。不竟要選擇的是潮流方面的取向還是運動性質的取向，現在的NIKE到底要朝哪一方面發展，老實說我還是處於完全不知道的狀態，而這也不是我能決定的。當然，如果NIKE公司內的誰來拜託我，我想我會接受這個工作的。這個意思就像我剛才說的，我認為adidas的潮流取向比較明顯，先不管adidas本身是怎麼認為的吧。」

——The adidas project you have been heavily involved in Originals by Originals (referred to hereon as ObyO) started in the spring summer season of 2009. How does it feel to be working on your third season for the project? Have you experienced any difficulties on the creativity side?
Kazuki: "I'm able to design quite freely in the apparel arena, but to be honest, it's more complicated when it comes to shoes. I'm sure other brands are the same but the reality is there are certain difficulties when it comes to developing a new outer sole."
——Is that for cost reasons?
Kazuki: "That's right. It costs more to develop a new mold. So what I work mostly on is designing uppers to go with the existing outer soles."
——Hiroshi, is it the same for you working with Nike?
Hiroshi: "No, that's not necessarily the case. There are certainly occasions I will just design the upper, but on other occasions I will design both the outer sole and the upper, but it takes time, of course."
Kazuki: "In my case I never design a shoe from the outer sole, so I guess it doesn't take as much time."
Hiroshi: "Having said that, it takes a certain amount of time from the planning stages to the point when the shoe actually goes on sale though, right?"
Kazuki: "Yeah, it takes more than a year."
Hiroshi: "I wish the process was faster."
——Hiroshi, what do you think of the products Kuraishi designs for ObyO?
Hiroshi: "Adidas is originally a sports brand, but compared to Nike and other brands, Adidas has a stronger fashion image. So from where I'm sitting, I'm inclined to think you have it a bit easier but what's the reality?"
Kazuki: "Mmm.. it's not so tough I guess but, not everything goes smoothly...."
Hiroshi: "I get the impression there would be a lot of people working at the top at Adidas who can understand your work..."
——Kazuki works on apparel and shoes for ObyO but Hiroshi, you mainly work with shoes, right?
Hiroshi: "Yeah. I only provide some ideas on apparel."
——Have you thought about going all out with Nike and designing apparel for them as well?
Hiroshi: "That would be difficult, considering Nike's current positioning in the market. I wouldn't know whether to take the product in a fashion direction or more towards sport. To be honest I'm not sure on Nike's direction and it's not something I can decide for Nike. Of course if Nike approached me I would do it. In that sense as I said earlier regardless of what Adidas would say about themselves, Adidas seems to have a stronger fashion element."

——我覺得實際上adidas比NIKE更早一步將「潮流」和「運動」明確區分。雖然現在NIKE有「NIKE」和「NIKE SPORTSWEAR」，不過adidas從前就有「adidas」和「adidas Originals」，並且名字和Logo都不一樣。

藤原「adidas應該從以前就對於時尚分野的行銷很有計劃的。這就是以歐洲為主的企業和以美國為主的企業不同之處。因為歐洲人比較能夠將自然和時尚結合。」
倉石「的確，這麼說的話或許真的是因為這樣呢。」

——無論是NIKE或adidas，他們都是提供最好的商品給運動員，本質是相同的。但是，不管哪一個品牌現在都想追求潮流的本質，所以才會利用藤原先生和倉石先生這類的創作者來達成目標吧。

藤原「應該是這樣吧，像我的話，與其說擁有時尚潮流的本質，不如說我是個具影響力的人比較適合。」
倉石「我以前開始穿SUPER STAR也是因為浩君，所以說浩君的影響力非常大哦。我成長的年代除了受到RUN DMC或BEASTIE BOYS這些海外樂團的影響外，還有以浩君為首的一些日本人也給我很大的影響。」

——那麼由倉石先生的觀點來看，藤原先生和NIKE的互動反映了什麼？

倉石「雖然我不太清楚細節部分，但從剛才對話中我聯想到的是藤原先生的工作比較簡單啊（笑）。」
藤原「真的嗎？（笑）要說的話，我的工作還是滿簡單的，不過還是會遇到不能盡速完成的挫折感。」
倉石「嗯，我覺得每一個企業都有相同的狀況，沒有經歷任何挫折感地完成工作是很難的事啊。」

——我個人印象是倉石先生除了服飾外，對於球鞋創作也十分有創意，我是看了完成品後才這麼認為。

藤原「嗯。我也有這種感覺。」
倉石「是嗎？」

——例如SUPER STAR，您將adidas的三條線以車線的方式呈現，鞋舌的Logo也改變成全新的樣式，在許多細節部分都嘗試挑戰很多新鮮的東西。

倉石「嗯，這樣說或許是真的。可以和擁有決定權的人直接了當的說出想法，決策的速度確實快很多。」
藤原「一樹君是從什麼時候開始與adidas合作的呢？」
倉石「大約5年前。」

——倉石先生，adidas會對你有甚麼要求？

倉石「adidas也許很在意浩君及NIKE的動向，一邊在旁觀看，一邊想著

——I think in reality Adidas made a clear divide between sports and fashion before Nike did. Now Nike is divided into "Nike" and "Nike Sportswear", but Adidas separated, both by name and logo into "Adidas" and Adidas Originals" at an earlier stage.
Hiroshi: "Adidas's marketing in the fashion field has always been solid. Maybe it's something to do with the difference between companies that are based in Europe and companies that are based in the US. Europe tends to be more inclined to fashion."
Kazuki: "You could be right there."
——Both Nike and Adidas started off with the same idea to provide better products to athletes. But now, both brands are looking to incorporate fashion into their products to some extent. That's perhaps the reason they have approached designers such as yourselves. Would you agree?
Hiroshi: I'm not so sure. I think in my case they are more interested in my influence on the market than my fashion sense."
Kazuki: "Hiroshi was the reason I started wearing Superstars. I think Hiroshi is very influential in that way. People of our generation were largely influenced by artists from overseas such as Run DMC and the Beastie Boys, as well as Japanese people like Hiroshi. The two influences have kind of blurred together in a way."
——What are your thoughts on Hiroshi's work with Nike?
Kazuki: "I'm not sure of the details, but following on from our earlier conversation, I think Hiroshi has it easier than I do. (laugh)"
Hiroshi: "You think so? (laugh) I guess I have it fairly easy, but I am constantly frustrated that I'm not able to do the things I want as quickly as I'd like to."
Kazuki: "I'm sure it's the same at any company. It's not often you can proceed with your work free from frustration."
——Kazuki, personally speaking when I look at your products, the shoes and apparel you design have a very free sense of style about them.
Hiroshi: "Yeah, I agree."
Kazuki: "Really?"
——For example with Superstar the way you changed the three lines to a stitch, or completely changed the logo on the tongue. You seem to have a lot of challenging new ideas.
Kazuki: "Mmmm.. you could be right. I speak directly with the guy who makes the final cut, so decisions are made quickly."
Hiroshi: "When did you start working with Adidas again?"
Kazuki: "Around 5 years ago."
——What do you yourself think Adidas expects from you?
Kazuki: "I think probably Adidas heard about the work Hiroshi was doing with Nike, and thought they also wanted to try something new by starting

曾經聽取藤原浩的建議，
把SUPER STAR鞋頭做成黑色。

Hiroshi gives me design ideas,
like the black toed Superstar I made was from him.

我並不是給adidas建議，
我只是給倉石一樹一點幫忙。

I wasn't advising Adidas.
I was just helping out Kazuki.

adidas也應該以東京作為情報發信地，並且要做出同樣新鮮的東西來吧，然後他們就會來詢問我了。」

——合作機緣是倉石先生還在職於A BATHING APE時所創作的「SUPER APE STAR」對嗎？

倉石「是的。當時adidas和其他品牌合作的商品沒有現在這麼多。」
藤原「其實我在數年前SUPER STAR 35周年時，曾建議一樹君如果有黑色鞋頭的SUPER STAR也不錯。」

——啊！？

倉石「對，那時被浩君這麼一說，真的就做了黑色鞋頭的SUPER STAR。」

——藤原先生，這種私下建議好嗎？

藤原「所以到現在我才敢說出來啊（笑）。但是，我並不是給adidas建議，只是給一樹君一點幫忙……順便問一下一樹君，如果你現在穿NIKE會不會被罵啊？」
倉石「如果這麼做的話，他們應該會非常生氣吧（笑）。」
藤原「是嗎？我穿adidas去NIKE也沒有被罵，不過被挖苦一下啦。」

——不不，這麼做一定被罵吧（笑）

藤原「我沒有和他們簽約說不能穿adidas啊（笑）。不過這種競爭對

手的關係，對於高層來說另當別論，但是現場的工作人員沒有這種情形吧，在adidas工作的人和在NIKE工作的人也有可能一起吃飯。」
倉石「這種情況應該會有吧。」
藤原「當然不會有實質情報交換，不過會交流潮流的共通話題吧。」

——那麼，最後可以告訴我們關於ObyO今後的發展嗎？

倉石「關於ObyO，我們會保留ObyO的名稱，然後會嘗試找一些其他設計師繼續創作新商品。現在是由我、JEREMY SCOTT、BECKHAM 3個人負責，或許有一天我們3個人都會退下……嗯，ObyO從一開始便是以這個概念出發，現在正準備這項工作，我們已經詢問過幾位設計師了。」

——對於ObyO擁有各種風格的設計師，然後他們以不同的創意詮釋adidas是這個企劃最有趣的地方。

藤原「是啊！adidas對於外部的時尚工作者來說，或許是他們很想一起合作的品牌。不過另一方面，我認為還是不要過份依賴外部比較好，因為不先做好自身品牌魅力的商品，和外面的人一起合作也是沒有意義的。說起來，一樹君你不妨正式加入adidas成為職員吧！與其說我能成為NIKE的社長，一樹君當上adidas社長的機率還來得更高呢！」
倉石「沒有這種事啦（笑）。」

a Tokyo project of their own. I suspect that's why they approached me."
——It all started when you were still working at Ape on Super Ape Star, right?
Kazuki: "That's right. At the time Adidas wasn't working with other brands as much as they do now."
Hiroshi: "A few years ago, at the time of Superstar's 35th anniversary, I remember I suggested to Kazuki that it would be good to have a Superstar model with a black toe cap."
——Really?!
Kazuki: "That's right. After that comment from him, I went ahead and made it."
——Hiroshi, so you were a hidden advisor?
Hiroshi: "I couldn't have said anything before of course. (laugh) But I wasn't making the suggestion to Adidas, I was just advising Kazuki. By the way Kazuki, would you get into trouble if you wore Nike?"
Kazuki: "I would probably get into a lot of trouble. (laugh)"
Hiroshi: "Really? I don't think Nike would be upset if I went to their office wearing Adidas. They might make some sarcastic remarks mind you."
——They would definitely give you hassle! (laugh)
Hiroshi: "I haven't signed any contract saying I can't wear Adidas. (laugh) Executives might be a different story but I don't feel there is a lot of rivalry between the two companies on the field level. I reckon people working at Adidas and Nike go out for dinner together once in a while."
Kazuki: "I'm sure they do."
Hiroshi: "I'm sure they don't exchange intrinsic information, but there must be some people exchanging ideas about the latest trends."
——Lastly, can you tell us a bit more about the future of ObyO?
Kazuki: "ObyO will continue with the same name, but there will be some changes to the designers working on the project. At the moment, it's me, Jeremy Scott and David Beckham, but there's a chance one day all of a sudden the three of us will be replaced. That's how the story's been since we started on ObyO. In fact they've already started approaching a few people in preparation."
——ObyO is interesting because there have always been various designers working on the project, and they all contribute their own individual interpretations of Adidas.
Hiroshi: "Adidas is a brand that a lot of people in the fashion industry want to work on. On the other hand, I don't think they should rely too much on outside resources. There's no point in working with outsiders if they don't make attractive products by themselves. Kazuki, why don't you become a formal employee at Adidas? I think there's more chance of you becoming the CEO of Adidas, than there is of me becoming the CEO of Nike."
Kazuki: "I doubt that very much. (laugh)"

Kazuki Kuraishi
倉石一樹

生於1975年。除了以adidas的creative product manager身份與德國總公司合作過許多重要企劃案外，也以fragment design及個人名義給眾多知名時尚品牌如Levi's Fenom、NEIGHBORHOOD等品牌提供設計。另外，也曾經參與Ian Brown、Tommy Guerrero等CD專輯的封面設計，活躍於各個不同界別。

Born in 1975. As well as working on a number of big projects as Creative Product Manager at Adidas's headquarters in Germany, Kuraishi has also designed for fragment design and a range of other fashion brands such as Levi's Fenom and NEIGHBORHOOD as a freelance designer. He has also worked in a range of other areas, including designing album covers for Ian Brown and Tommy Guerrero.

AIRWALK
ENIGMA

1980年代後期至1990年代前期，是球鞋文化和滑板文化互相交融的黎明期。具體來說，滑板運動者開始穿著滑板鞋以外的球鞋，而非滑板運動者則開始穿著滑板鞋，正因如此，讓雙方都接受的便是AIRWALK的「ENIGMA」。除了必備滑板鞋的實用性外，麂皮材質的風格及藍色×黑色×米色的配色組合十分美麗，與當時的街頭潮流文化十分契合。「這雙球鞋很適合玩滑板時穿著，最重要的是出色的潮流感。」藤原浩也對這雙球鞋的功能性給予極高評價。

In the late 1980s and early 1990s a cultural cross-over between sneakers and skateboarding began, as skaters began to wear sneakers other than skate shoes and non-skaters began to wear skate shoes on the street. A shoe that made the hearts of both parties jump, was AIRWALK's ENIGMA. The suede texture and the blue x black x beige coloring was simply beautiful. The shoe performed well as a skate shoe while also epitomizing the street culture trends of the time. "This shoe was really comfortable to wear but above all, damn stylish." Hiroshi Fujiwara comments admiringly.

藤原浩約在1990年進行滑板運動時穿過的AIRWALK「ENIGMA」。

AIRWALK's ENIGMA that Hiroshi wore for skateboarding.

NORTHWAVE
ESPRESSO

「AIR MAX」風潮曾經發展成社會的話題現象，但在快要結束的1996～1997年間，便產生一股相反的力量，引發另一極大風潮的是以一般生產技術製作的球鞋。在這種狀況下爆發人氣的是曾經寫下紀錄的NORTHWAVE「ESPRESSO」。NORTHWAVE本身是製造滑雪板的運動品牌，球鞋的生產數量本身並不多，加上因為突然受到各界的矚目，供求的平衡在一瞬間崩壞，市場的庫存變得所剩無幾，使得價格不斷飆漲。究竟當時為什麼外形這麼笨重的厚底球鞋會突然造成爆炸性的人氣呢？雖然至今箇中的原因不能確定，但是大有可能是經過藤原浩在雜誌中推薦，也未嘗不是一種有力的幫助。

The AIR MAX boom that at one stage reached the status of social phenomenon, started to cool down around 1996-97, the repercussions of which, incurred a massive movement in low-tech sneakers. In such an age, one shoe that received phenomenal acclaim was NORTHWAVE's ESPRESSO. NORTHWAVE was originally a relatively low production Italian snowboard brand. The sudden explosive popularity of the ESPRESSO resulted in an imbalance of demand versus supply, which caused a sharp increase in price, with stock selling out everywhere. It is not known exactly why the unusual chunky shaped platform shoe became a huge hit, but it is said that Hiroshi Fujiwara contributed largely when he introduced the shoe in a magazine.

NORTHWAVE「ESPRESSO」曾經 在1990年 代 前期至後期風靡一時，藤原浩曾經也有一陣子非常愛穿這雙球鞋。「現在想起來會覺得有點不好意思呢（笑）。」

NORTHWAVE's ESPRESSO which dominated the mid to late 90s. Hiroshi wore the shoe habitually for a period. "I'm slightly embarrassed when I think back on it now. (laugh)"

VANS
CLASSIC SKOOL / CHUKKA

自1966年誕生至今，VANS一直支持著眾多滑板運動的愛好者。藤原浩在1970年代當時也曾經穿過，除了玩滑板之外日常生活中也會穿著。VANS除了有被稱為HI-TOP的高筒「SK8-HI」及JAZZ的「OLD SKOOL」外，還有眾多各式各樣的鞋款，而藤原浩特別喜愛的是款式設計較簡單的「CLASSIC SKOOL」及中筒的「CHUKKA」。另外，VANS的圖案設計素來多以格紋或花紋等有趣的元素為賣點，但是相比之下，藤原浩還是喜歡穿著單一色系的簡約鞋款。

VANS is a brand that has received support from many skaters from the late 70's to the present. Hiroshi Fujiwara also wore VANs not only for skateboarding but also for everyday use during the 1970s. The brand released various models including the SK8-HI which was often referred to as "High-Top", and the OLD SKOOL also known as the "Jazz". Hiroshi particularly loved the simply designed CLASSIC SKOOL and middle cut Chukka type. Unique patterns such as checkered flag and flower print are characteristic of VANS shoes, but Hiroshi preferred the more simple versions.

款式設計簡單的「CLASSIC SKOOL」採用帆布材質，圓弧形的楦頭是其特色，分別有高筒和低筒款式。

Simply designed CLASSIC SKOOL using canvas. The rounded toe cap makes the design unique. The shoe came both in high-cut and low-cut.

以具有潮流感的街頭鞋款獲得滑板運動者以及一般人士支持的「CHUKKA」，從鞋子側面破損狀態來看，可以觀察出這是藤原浩經常玩滑板運動時所造成的。

This is a Chukka model, which was supported not only by skaters but also by many people as stylish street shoes. You can see by the way the shoes have worn thin at the side that Hiroshi used to wear habitually for skateboarding.

PRO-Keds
ROYAL PLUS

PRO-Keds的代表鞋款「ROYAL PLUS」，跟很多其他球鞋一樣，也是經藤原浩影響下造成的流行球鞋。這款球鞋原本是為籃球運動而設計及開發，藤原浩在中學時代隸屬籃球隊的時候也有穿過。「在那個時代，無論是什麼球鞋品牌，只要是來自海外的品牌出品，就會讓人覺得非常珍貴。我自己也是在《POPEYE》雜誌的影響下成長的，所以對於美國品牌的憧憬特別強烈。」另外，在衆多「ROYAL PLUS」的版本中，是以哥倫比亞製造的最有名氣，但不幸是PRO-Keds的哥倫比亞工廠在1996年關閉了，而在此之前生產的鞋款也被球鞋界稱之為「最後的哥倫比亞」，也因此在球鞋迷之間曾經引起非常激烈的爭奪戰。

The ROYAL PLUS from PRO-KEDS are another shoe also considered to have brought to popularity by Hiroshi. This model was originally developed as a basketball shoe. Hiroshi wore the shoes for basketball club when he was a junior high school student. "At that time, sneakers from overseas had a novelty to them regardless of the brand. My generation grew up reading "Popeye" magazine, so I had a strong yearning for American brands." ROYAL PLUS's produced in Colombia were especially famous, but PRO-KEDS closed their plant in Colombia in 1996. Models sold before that time are referred to as "Last Colombia". There is strong competition among enthusiastic fans to get these shoes."

藤原浩在1980年代主要的愛用鞋款，是來自哥倫比亞製造的「ROYAL PLUS」。以前的鞋款才能呈現出麂皮獨有的味道及風格，十分美麗。

ROYAL PLUS made in Colombia that Hiroshi Fujiwara wore in the 1980's. It has that beautiful suede texture you can only find in sneakers of the past.

asics
GELMAI

藤原浩所擁有唯一的asics是「GELMAI」。在鞋款多為古典基本樣式的asics中，這雙球鞋可以說非常前衛，仔細觀察細部設計，偏向外側的鞋帶和網狀材質的鞋面等，都有點像NIKE的「FOOTSCAPE」（P.52－53）。藤原浩入手這雙球鞋是在1999年的時候，而這也是在偶然的情況遇到的。「當時我住在碑文谷附近，這雙鞋是在家裡附近的DAIEI發現的，asics居然有推出這種鞋款真是讓我十分驚訝！那時候的心情，就像自己發現了沒人注意的球鞋，真的很開心呢！」他就是這樣敘述與這雙球鞋的相遇經過。

The GELMAI is the only ASICS shoe that Hiroshi Fujiwara owns. It's a pretty outrageous shoe for ASICS who are well-known for their more classical designs. When you observe the details of this shoe, the shoelaces are positioned further apart and the mesh upper and other elements are reminiscent of NIKE's FOOTSCAPE (P.56-57). Hiroshi bought this model in 1999. "I lived in Himonya at the time, and found this model at the neighbourhood discount store. I was amazed that ASICS made such a shoe.", Hiroshi recalls on his encounter with the shoe. "I still like discovering sneakers that nobody else has paid much attention to."

藤原浩在1999年入手的「GELMAI」。在碑文谷DAIEI偶然發現黑色款式，因而開始尋找其他顏色版本。

ASICS GELMAI that Hiroshi bought in 1999. After he found the model in black at a discount store in Himonya, he began to look for it in other colors.

UNDERCOVER
Original Sneaker

藤原浩愛用的球鞋以運動品牌居多，但是有時候還是會穿著時尚潮流品牌推出的鞋款。這雙UNDERCOVER的球鞋便是其中之一，3雙中最喜歡穿的是黑色那款。「UNDERCOVER擁有很強的品牌信念，能將球鞋融入時尚潮流中。」

Most sneakers Hiroshi likes to wear are sports shoes, but sometimes he wears original shoes from fashion companies. This UNDERCOVER sneaker is an example. He has three pairs of the model, but he likes to wear the shoe in black. "UNDERCOVER have integrated sneakers into fashion with quite some conviction."

於1997年推出，以CONVERSE「JACK PURCELL」的設計為概念，鞋頭變得更有份量感，並在「SMILE」的位置加入品牌的名稱。

Released in 1997. In the motif of the CONVERSE JACK PURCELL design, this shoe has a bulkier toe than the original. The brand's name appears on the so-called "smile" part of the shoe.

Reebok
INSTAPUMP FURY

1989年誕生的Reebok「PUMP」，是將空氣充填至鞋面內的技術，再以這股壓力支撐足部，在當時來說是劃時代的新設計。要說主流的鞋款則為1994年上市的「INSTAPUMP FURY」。「我覺得『PUMP』這個系統是很有趣的設計，實際穿起來也很舒適，但是有時候也會令人覺得，日常生活中穿著時有必要如此高度的包腳感覺嗎？」

The Reebok PUMP was introduced in 1989. It pumps air into the upper, building pressure to support the foot. This technology was a revolutionary system at the time. The shoe that boosted this technology into the mainstream was the PUMP FURY, released in 1994. "I think the Pump is an interesting system and comfortable to wear, but I wonder, does a shoe for daily wear really require such technology?"

2001年在CHANEL服裝秀中使用的特別款式，只特別分配給一小部分的相關業者，不作一般銷售。

This is a special model used in the Chanel fashion show in 2001. The shoe was given out to a handful of people involved in the event. Not for general sale.

Reebok
BOKS

Reebok在1994年左右推出休閒系列「BOKS」的「Hard Drive」鞋款。運動品牌推出這類休閒鞋款是比較罕有的，這似乎也被藤原浩高感度感應的雷達搜尋到了。

This is the HARD DRIVE from the casual line, BOKS. Released by Reebok in 1994. It was unusual for a sports brand company to release this kind of casual shoe. But it seems that the model showed up on Hiroshi's high-powered radar.

Teva
Response 2

Timberland
Sandal

Sandal

藤原浩的涼鞋選擇

基本上平日大家都是穿著球鞋，不過在炎熱的夏天也會偶爾想換上戶外涼鞋輕鬆一下吧，藤原浩也有著相同的想法。在這裡介紹的是，藤原浩過往曾經穿過的4雙不同品牌的戶外涼鞋。最古老的一雙，是1990年代前期他最愛用的Teva「Response 2」。當時，以混色厚襪搭配Teva或Timberland等戶外休閒品牌涼鞋的造型非常流行，而藤原浩當時也早先一步實踐了這種搭配方式。另外兩雙，「AIR DESCHUTZ」和「JORDAN TRUNNER」皆為NIKE的出品，藤原浩先後在1990年代後期至2000年前後穿過。「『AIR DESCHUTZ』是第一雙中底配備AIR的acg戶外涼鞋，所以穿起來非常舒適。另外，『JORDAN TRUNNER』的外觀雖然十分獨特，極具設計感覺，但是穿起來的卻不是最好穿的一雙。（笑）」

Sneakers can be worn for everyday life, just as sandals are for relaxing on a hot summer day. Hiroshi also shares this thought. We picked up four pairs of sandals that Hiroshi has worn in the past. The oldest is the RESPONSE from TEVA, which he wore most in the early 1990s. At that time, wearing outdoor sandals such as TEVA and TIMBERLAND with socks became a trend. Hiroshi quickly incorporated that style into his fashion. The AIR DESCHUTZ and the JORDAN TRUNNER were released by Nike, and Hiroshi wore them from the late 1990s to around 2000. He comments, "The DESCHUTZ was the first sandal with an air sole unit and it was very comfortable to wear. The JORDAN TRUNNER has a unique look but it wasn't the most comfortable shoe. (laugh)"

NIKE
AIR DESCHUTZ

NIKE
JORDAN TRUNNER

Dialogue
Hiroshi Fujiwara × Takashi Imai (MADFOOT!)

對談 藤原浩 × 今井崇（**MADFOOT!**）

相隔數年再相見，同樣熱愛「球鞋」和「音樂」的
MADFOOT!今井崇和藤原浩以下將會為我們聊到
東京球鞋文化的過去、現在和未來。

Hiroshi Fujiwara and Takashi Imai of MADFOOT! Re-unite for the first time in years.
Both share a passion and history in sneakers and music.
They discuss the past, present and future of Tokyo Sneaker culture.

藤原「和今井君初次見面是在GAS BOYS的時候，當時你是在球鞋店工作對吧？」

今井「是的。我一邊在上野的『KANEOKA』上班，一邊進行音樂活動。說到GAS BOYS，第一張唱片專輯封面上的文案還是請浩君寫的呢。」

藤原「你一說讓我想起真有這回事。雖然我已經不記得寫了些什麼……」

——這大概是什麼時候的事情呢？

今井「我記得大概是1991或1992年左右。」

——原來兩位最初的接觸點不是球鞋而是音樂啊？

藤原「一開始是這樣的。但是當時我們周遭做音樂的人，每個都喜歡球鞋。那麼，今井君你當時穿的是什麼鞋款呢？」

今井「我常穿的是adidas的SUPER STAR或CAMPUS。」

藤原「果然還是跟隨Beastie Boys等外國藝人的潮流。或許當時的時代可說是adidas派多過NIKE派。」

今井「不過浩君在當時常穿的是DUNK吧。穿著深藍×黃的DUNK玩滑板的身影我還留下深刻印象。」

藤原「嗯，沒錯，當時真的很常穿DUNK。今井君從那個時候開始就一直待在球鞋業界嗎？」

今井「是的。我在這個業界已經將近20年了。」

藤原「上野之後去了原宿的『atmos』工作吧。」

今井「是的，沒錯。」

藤原「『atmos』這個名字是我取的。」

——啊！？真的嗎？

藤原「嗯。LOGO也是我設計的。本明君來找我說『我要開新的店舖，可幫我一起想個店名嗎？』於是我就幫他了。免費的（笑）。」

——我第一次聽說，今井先生知道這件事情嗎？

今井「是的，我知道哦。而我進入『atmos』是在開幕後2個月左右，然後待了大約2年吧」

——『atmos』的概念也是藤原先生想的嗎？

藤原「是的。免費的哦（笑）。不過，像『atmos』這樣的提案式球鞋店舖在當時真的很稀奇。雖然我不知道到底有沒有獲利。」

今井「這方面倒是還不錯哦。當時本明君以魔鬼般的下訂方式大量進貨，所以我也一直為了庫存的分類很辛苦（笑）。」

藤原「這麼多的庫存都保管在哪裡呢？」

今井「當時有3個倉庫，倉庫裡都是堆積成山的球鞋，每次進貨我都得在球鞋的大海中游泳（笑）！」

藤原「那麼，在『atmos』待了2年，之後呢？」

今井「一邊在『atmos』工作，一邊創立了MADFOOT!。」

——MADFOOT!是在什麼時候創立的呢？

今井「2001年。」

藤原「那也將近10年了啊！做到現在的感覺如何？我認為以個人的力量創立球鞋品牌應該非常辛苦吧，一個人可以做到嗎？」

今井「不，做不到啊（笑）。剛開始的時候情況是不錯的。當時，包括visvim的中村君等，不是許多人都創立了新的品牌嗎？就像國產球鞋潮流的

Hiroshi: "The first time I met you Imai, was back in the Gas Boys days, wasn't it? Remind me, had you already started in the sneaker business at that point?"

Imai: "Yes, I was working at a sneaker shop in Ueno, Tokyo called "Kaneoka", while keeping up my music career. Speaking of Gas Boys, we got you to write the liner notes for the first album we released, didn't we."

Hiroshi: "That's right. I forget what I wrote though."

——When was that?

Imai: "I think it was around 1991-92."

——So your first connection was not through sneakers, but through music then?

Hiroshi: "Originally, yes. But the people we associated with at the time without exception were all into sneakers as well as music. Imai, what were you wearing at the time?"

Imai: "Adidas Superstars, Campus..."

Hiroshi: "Sneakers worn by foreign music artists like the Beastie Boys. At the time there were perhaps more people wearing Adidas than Nike."

Imai: "But you were wearing Dunks. I can still clearly remember you skateboarding in those navy x yellow Dunks even now."

Hiroshi: "Yeah, I wore those a lot. Have you been in the sneaker industry since then?"

Imai: "Yeah, I've been in the industry for twenty years now."

Hiroshi: "After Ueno, if I remember correctly you went to Atmos in Harajuku, right?"

Imai: "That's right."

Hiroshi: "Atmos. I was the parent that named that store."

——What?! Really?!

Hiroshi: "Yeah. I made the logo too. Honmyo told me he was opening a new store and confided in me about a name for it. So I came up with the name and logo, free of charge! (laugh)

——That's news to me. Imai, did you know that?

Imai: "Yeah, I knew about it. I started working at Atmos two months after they opened. I was there for about two years."

——Hiroshi, did you come up with the concept for the store as well?

Hiroshi: Yeah, I did that for free too. (laugh) It was an unusual proposal for a sneaker shop at the time. I never heard if it ended up profitable or not"

Imai: "The store did good business. Honmyo was making monster orders and bringing an absurd amount of stock into the store. It was me who had the tough task of sorting all the stock out. (laugh)"

Hiroshi: "Where were you storing all that stock?"

Imai: "We kept the stock in about three different places. In each, there was a mountain of sneakers I had to swim through every time we got a new order in. (laugh)"

Hiroshi: "Where did you go after Atmos?"

Imai: "I was preparing for the launch of MADFOOT! while I was working at Atmos."

——When was MADFOOT! established?

Imai: "2001."

Hiroshi: "So it's been ten years then. How has it been so far? Running a sneaker business on your own sounds tough but is it doable?"

Imai: "Nope. Not possible. (laugh) I was in good shape at the beginning. There were a lot of people at the time, including Nakamura from visvim, starting their own brands. There was a boom in the domestic sneaker industry, and we were getting a lot of attention from the media. Any shoes I made with MADFOOT! were a sure sale. But it didn't last forever. After that, it became more and more

形式,媒體也經常報導。MADFOOT!也是只要推出商品就能熱賣的狀態,但是,結果也是有一時性的部分,在那之後的資金方面就變成一個人無法應付的嚴苛狀態了。」

藤原「原來有這個時期啊?」

今井「結果球鞋的宿命還是必須大量生產吧。有時候工廠會因為沒有一定的生產數量而不接受訂單。」

藤原「對啊,製造球鞋在生產方面是很辛苦的。」

今井「剛好在這樣的狀況下,伊藤忠來找我合作。」

藤原「小型的個人球鞋品牌和大企業合作,今井君應該是第一個吧。」

今井「但是,老實說當時真的因為資金問題非常辛苦,以自己的資本能力來說還是有限。本來是因為想要創作商品而開始,但是結果卻好像不停地因為資金問題而困擾。」

藤原「實際上在商品推出前就需要花錢,而在商品販售後到入帳這段時間又有時間上的間隔。」

——注入了企業資金後,MADFOOT!的運作比之前更順暢了嗎?

今井「這是無庸置疑的。還可以赴國外參加展示會,這是光靠自己的資本所不容易達到的事情。」

藤原「那麼,某種意思是和伊藤忠合

作可以在沒有精神負擔的狀況下創作。」

今井「對啊,現在真的是沒有精神負擔。」

藤原「現在也是今井君自己直接和工廠溝通嗎?」

今井「是啊。我常常出差到中國大陸的工廠。」

藤原「市場行銷呢?也是自己做嗎?球鞋在某種程度上是必須以相同的形式繼續銷售,這點讓人感覺不容易,相反地說或許又輕鬆。」

今井「在我來說是滿輕鬆的。不,可以說是輕鬆嗎,例如服飾品牌在春夏和秋冬季節時會將全部的商品更換,那真的是大工程。如果是球鞋店要在換季時將庫存全部更換,將會是非常困難的事情。」

藤原「我認為這就是製作球鞋困難的地方。製作相同的商品是很有效率的,但是顧客總是想追求新的東西吧。」

今井「如果相同商品的銷售可以持續創造出好成績就好了,例如NIKE AIR FORCE 1、adidas SUPER STAR 或是CONVERSE ALL STAR。」

——MADFOOT!的客人以老顧客居多?還是新客人比較多呢?

今井「和伊藤忠合作之後,市場的範

---

difficult to handle the business on my own, financially speaking as well."

Hiroshi: "Sounds like you had it tough for a while."

Imai: "Sneakers have to be mass produced, or it just doesn't work out. If you don't make a certain amount, the factory just won't take the work."

Hiroshi: "Yeah, that's the tough thing about sneakers."

Imai: "It was around that time that Itochu contacted me."

Hiroshi: You were the first indie sneaker brand to join forces with a major trading company, weren't you?

Imai: "The truth is I was going through a really hard time on the financial front. It didn't matter what I did, there was a limit as to how far my personal funds could take me. I started in the business because I wanted to design sneakers, but cash-flow problems were taking their toll."

Hiroshi: "You need considerable finances to launch a product, and it takes time before the product sells and the money actually starts to come in."

——Did it become easier to work on MADFOOT! when the capital came in from the trading company?

Imai: "Definitely. I was able to start showing our products in exhibitions overseas for a start. Something that would not have been possible to do on my own funds."

Hiroshi: "So since Itochu came on board, you're able to work without the extra stress then?"

Imai: "That's true. I definitely have a lot less stress now."

Hiroshi: "Are you still in direct communication with the factories?"

Imai: "Yes. I visit the factories we have contracted with Itochu on a regular basis. I have staff in China that I travel with in China, Taiwan etc."

Hiroshi: "What about marketing? Do you do that yourself too? I think with sneakers you have to continue selling the same models for a slightly longer period of time. Do you find that a challenge or easier?"

Imai: "It's easier that way for me. What I want to say is, I think it's more amazing how clothing brands for example manage to change their entire lineup from season to season. It would be hard for a shoe brand to replace their entire stock every season."

Hiroshi: "I tend to think that's what's difficult about making shoes. Maybe it's more efficient to be making the same products, but customers are always after something new."

Imai: "What you need is to create a product that will continue to sell for you. For example, Nike Airforce 1, Adidas Superstar, Converse Allstar etc."

——Do MADFOOT! customers consist more of repeat buyers, or new

---

在日本國內的發展已到極限了,
接下來想擴展到海外的市場。

There's a limit to how well I can do in the domestic market,
so from here on I'd like to focus also on the overseas markets.

圍擴大了，我覺得最近以新的顧客比較多吧。」

藤原「現在都在什麼樣的店舖販賣呢？」

今井「主要以『ASBEE』為主。在東京也有路面店，其他縣市則多在AEON購物中心內。」

藤原「會依照店舖不同而區分銷售的商品款式嗎？」

今井「我是想要這麼做，不過實際上還沒有。」

——從去年開始MADFOOT!在女性顧客中獲得極高的人氣呢。

藤原「真的，常常在街上看到女生穿著鮮豔配色的MADFOOT!。」

——甚至在這之中有些女生以為MADFOOT!是女性品牌，在我們看來是很不可思議的。說到這個，之前今井先生在接受《SHOES MASTER》訪問時曾經說過，自己並不在意品牌形象，創作的商品如何被詮釋也無所謂，而是將最終的判斷交給顧客。這種心情到現在也沒有改變嗎？

今井「是的。」

藤原「關於這點我也很能了解。音樂也是相同的，在曲子完成時已經從自己的東西變成別人的東西了。」

今井「我了解。我在做音樂的時候也有這種感覺。」

藤原「那麼，現在MADFOOT!的設計師只有今井君一個人嗎？」

今井「對，只有我」

藤原「不會很辛苦嗎？可以再增加人手啊，例如年輕一輩的設計師。」

今井「如果這樣的話我也可以輕鬆一點……但是，我覺得一個人做有時候比較能輕鬆的改變工作方向。例如去年熱賣的是女生用的鮮豔鞋款，不過這只是一時性的潮流，退潮的速度也很快。如果可以輕易改變工作內容，也就可以很快的應對市場環境的變化。」

藤原「原來如此，或許是的。那麼，接下來的MADFOOT!將如何發展呢？」

今井「在日本國內的發展已經極限了，接下來想擴展至海外的市場。」

藤原「中國大陸呢？」

今井「已經在做了哦，大約1年前開始的。在歐洲的展示會也會依照計劃出展，今後也預計到美國參加展示會。」

藤原「附帶問一下，如果想利用既有的鞋型，以新的配色或圖案創作新商品銷售，大概要花多少時間？」

今井「大約半年後吧。」

藤原「半年後就可以有商品銷售了？很快耶！」

customers?

Imai: "Since I teamed up with Itochu, my market has expanded, so recently the majority of our customers are first-time buyers."

Hiroshi: "Which stores are you selling through at the moment?"

Imai: "Mainly ASBEE. We have some other dealers in Tokyo. In rural areas, our products can be found in shopping centers and AEON stores."

Hiroshi: "Are the shoes on sale different depending on the store?"

Imai: "I'd like to select products for stores, but the fact is I haven't been able to yet."

——MADFOOT! has suddenly become very popular among ladies since last year.

Hiroshi: "That's true. I've seen a lot of girls wearing brightly coloured MADFOOT! shoes on the street recently."

——There are some girls who actually think MADFOOT! is a ladies brand. Inconceivable to the likes of us. Speaking of which, in a previous interview for SHOES MASTER, you stated that you didn't care about the image of your brand or how your products were received, and that you preferred to leave the final judgment up to the customer. Your opinion hasn't changed since then?

Imai: "No change."

Hiroshi: "I feel the same way. It's the same as music. Once a song has been produced, the song no longer belongs to you, but to the person who is listening to it."

Imai: "I can understand that. I felt the same way back when I was making music."

Hiroshi: "Are you still the only designer for MADFOOT!?"

Imai: "Yes. Just me."

Hiroshi: "Isn't that hard? You should get someone in. A young designer say."

Imai: "It would make things a bit easier...But in a way, it's easier for me to go from idea to idea working on my own. For example last year brightly colored ladies shoes were all the rage, but it was a temporary trend. The tides change fast. The faster I'm able to work, the faster I can respond to such changes."

Hiroshi: "I see what you're saying. What are your plans for MADFOOT! now?"

Imai: "There's a limit to how well I can do in the domestic market, so from here on I'd like to focus also on the overseas markets."

Hiroshi: "What about China?"

Imai: "I'm already in the China market. We started displaying our shoes in China about a year ago. We will also be displaying our products at trade shows in Europe as we have been doing, and would like to start doing the

今井君，下次試用馬德拉斯格紋
設計一雙球鞋吧。

Imai, you should design a pair of sneakers in madras check!

——藤原先生在NIKE如果想創作商品的話，大概要花多少時間呢？

藤原「如果利用既有的鞋型就比較快，但是半年是有困難的，最少要1年的時間。」

今井「不過如果是NIKE的話，有些鞋款不需要花到2年的時間嗎？」

藤原「要哦，但是最近有稍微快一點。」

——例如，如果要請藤原先生設計一款新配色的TENNIS CLASSIC，從提案到上市銷售大約需要多少時間呢？

藤原「這是不需要太長時間的，但是在前一個階段，也就是推薦TENNIS CLASSIC的時候，反而是要花很長的時間。舉個例子，若當時熱賣的鞋款是AIR FORCE 1，那麼為什麼熱賣的是AIR FORCE 1但一定要推出TENNIS CLASSIC呢，加起來大概花了4年的時間吧。」

今井「啊，要這麼久啊！」

藤原「甚至於剛開始時還說絕對不能將SWOOSH以打洞的設計方式出現呢。」

今井「浩君有試過從頭開始設計一雙新球鞋的經驗嗎？」

藤原「偶爾有這種案子，但是比較多的情況是從以前的球鞋中發現有趣的款式，然後提出再上市的企劃案。」

——關於藤原先生所經手設計的球鞋，今井先生有什麼感想？

今井「我覺得非常有品味，無論哪一款都很有品味。」

藤原「很樸素吧（笑）。」

今井「對於AIR FORCE 1的詮釋方法也很獨特，特別是HTM的設計非常嶄新哦。將AIR FORCE 1利用上等材質搭配出時髦樣式的設計手法，在當時是不曾看過的。」

藤原「因為在某種意義上，我是『進攻隊長』的角色，像我這種以時尚界的操作方式和NIKE合作的公司外部人員，在當時可以說是完全沒有的……。對了，說到這個，剛剛在電視上看到冬季奧運會，美國的滑雪板選手所穿的BURTON馬德拉斯格紋外套很好看哦！今井君，下次可以試試採用馬德拉斯格紋作材質，設計出一雙球鞋吧。」

今井「哦，我馬上就可以做哦。可以用HF（藤原浩）推薦款的方式來銷售嗎？（笑）」

藤原「那我會被NIKE罵，我偷偷的口耳相傳給你就好了。（笑）」

same in the Americas next."

Hiroshi: If you wanted to reproduce a model you already had in a new colour or pattern now, how long would it take?"

Imai: "Around 6 months I guess."

Hiroshi: "You could get the shoe on sale in 6 months? That's fast."

——How long does it take for you with Nike?

Hiroshi: "It would be faster if it was an existing model, but still 6 months would be difficult. It would take at least a year."

Imai: "For Nike, depending on the model, wouldn't it take closer to two years?"

Hiroshi: "Yeah. It's gotten a bit faster recently though."

——For example, if you came up with the idea to bring Tennis Classic out in a new colour, how long would it take from the proposal to the time the new colour goes on sale?

Hiroshi: "It wouldn't take that long. The problem is the stage before that: the decision to go to work on Tennis Classic. For example, if Airforce 1 was selling at the time, Nike would want to focus their resources on Airforce 1. It took four years for that colour to come out for Tennis Classic."

Imai: "That's a long time."

Hiroshi: "When I first started collaborating with them, they told me punching on the Swoosh was completely out of the question. So I guess I've come a way with them since then"

Imai: "Do you make shoes from scratch sometimes too?"

Hiroshi: "Sometimes, but more often than that, I look for interesting shoes from the past, and make suggestions on how they might be brought back into the market."

——Imai, what do you make of the shoes that Hiroshi has worked on until now?

Imai: "I think they are all very sophisticated. They all have class."

Hiroshi: "If not a little low-key. (laugh)"

Imai: "I thought his interpretation of Airforce 1 was extremely unique. HTM certainly a breath of fresh air too. His approach with Airforce 1, to dress it up in high-quality leather, was particularly unique for the time."

Hiroshi: "In a sense it was new waters for Nike as well. At the time there were no other third parties from the fashion world working together with Nike. By the way, I was watching the Olympics earlier and the madras check outer the US snowboarder was wearing was so cool. Imai, you should design a pair of sneakers in madras check!"

Imai: "I'm on it! Can I say the model was endorsed by HF?! (laugh)"

Hiroshi: "I will get into trouble from Nike, so don't let it get on paper. (laugh)"

Takashi Imai
今井崇

MADFOOT!統籌。1990年代除了在球鞋店舖工
作，也以饒舌團體「GAS BOYS」的MC身份活
躍於音樂界。之後經歷「atmos」的統籌後，在
2001年創立MADFOOT!。

MADFOOT! Director. During the 1990s, Imai
flourished as MC of the legendary group, Gas Boys,
while working for a sneaker store. After that, he
became director of the Harajuku store, Atmos, and
in 2001 established the MADFOOT! brand.

Timeline of Hiroshi Fujiwara and Tokyo Sneaker History

藤原浩與東京球鞋歷史年表

以藤原浩及周邊發生的事為主軸，
將1980年代至2000年代的東京球鞋文化歷史以時間表形式整合。

With a focus on Hiroshi Fujiwara and what was happening around him,
we have summarized Tokyo Sneaker Culture History chronologically from the 1980s to the 2000s.

1980

1980年代前期　流行美國西岸風格
Early 1980s: American West Side

1980年代前期～中期　流行DC品牌
Early to Mid 1980s: DC

1982

NIKE推出「AIR FORCE 1」
Nike releases the AIR FORCE 1.

藤原浩至倫敦遊學。
和Vivienne Westwood及Malcolm McLaren建立了深厚交情

Hiroshi Fujiwara goes to London to study.
He deepens a friendship with Vivienne Westwood
and Malcolm McLaren.

Reebok推出健康舞鞋款「FREESTYLE」。
健康體操運動風潮席捲全球
Reebok releases Aerobic shoe, the FREESTYLE.
A Fitness boom takes place.

1984

洛杉磯奧運開幕。
穿著NIKE的Carl Lewis獲得4項冠軍
Los Angeles Olympics is held.
Carl Lewis wins four gold medals wearing Nike shoes.

1985

NIKE推出「DUNK」
Nike releases the DUNK.

New Balance推出「1300」。
當時在日本的售價為39,000日圓
New Balance releases the1300.
The price at that time was 39,000 yen.

藤原浩和高木完組成「TINY PANX」
Hiroshi Fujiwara forms band Tiny Panx with Kan Takagi.

NIKE和NBA超級球星Michael Jordan簽約。
同時推出「AIR JORDAN 1」
Nike signs NBA super-rookie Michael Jordan.
The AIR JORDAN 1 is released.

VANS_ CLASSIC SKOOL

adidas_CAMPUS

PUMA_CLYDE

adidas_SUPER STAR

1980's

對1980年代的藤原浩來說，「是否適合滑板運動時穿著」是他選擇球鞋時的一大要素。雖然如此，但機能性以外，潮流性和設計也是不能遺忘的重點。

What are they like to skate in? That was the main criteria for selection of sneakers for Hiroshi Fujiwara in the 1980s.
But he didn't only care about the shoe's functionality, but also its fashion relevant aesthetic and design.

1989

1980年代中期～後期　流行澀谷街頭＆美式休閒造型

Mid to Late 1980s: Shibuya Casual and American Casual

1986	1987	1988	1989
AIRWALK公司創立 AIRWALK is established.	NIKE推出「COURT FORCE」 Nike releases the COURT FORCE.	New Balance推出「576」、「996」 New Balance releases the 576 and 996.	NIKE推出戶外休閒系列「ACG」 Nike launches an outdoor product line named ACG.
雜誌《Boon》創刊 Magazine "Boon" publishes its first issue.			由Spike Lee執導的電影《Do the Right Thing》上映。 劇中人物多數穿著 AIR JORDAN "Do the Right Thing" directed by Spike Lee is released. Many characters in the movie wore the AIR JORDAN.
RUN D.M.C〈Walk This Way〉大流行，adidas成為風潮 RUN D.M.C "Walk This Way" becomes a big hit. Adidas boom.			
Beastie Boys發行首張專輯《Licensed To III》 Beastie Boys releases their first album "Licensed To III".			

NIKE_DUNK

NIKE_AIR JORDAN 1

COURT FORCE

AIRWALK_ENIGMA

在廣大領域中一直豎起高感度的天線，經常提供周遭最新情報的藤原浩受到媒體的矚目。
特別是在球鞋界別方面，他以銳利眼光發揮所長，因為他穿過而引起風潮的球鞋數量難以估計。

Media focus in on Hiroshi Fujiwara who put his antenna out in a wide range of fields and provided fresh information to those around him.
In particular, Hiroshi exhibited his excellent sense of style in the sneaker industry.
There are countless pairs of sneakers that subsequently became popular after he wore them.

1990

1990年代前期～中期　二手衣物&復古風格成為主流

Early to Mid 1990s: Second-hand/Vintage

1990	1991	1992	1993	1994
NIKE推出「AIR JORDAN 5」 Nike releases the AIR JORDAN 5.	PUMA發表最新FIT SYSTEM「DISC SYSTEM」 Puma introduces new fitting system named the DISC SYSTEM.	巴塞隆納奧運「夢幻球隊」大活躍。在日本興起NBA風潮 The US "Dream team" show a spectacular performance at the Barcelona Olympics. An NBA boom starts in Japan.	「NOWHERE」於原宿開幕 The store "NOWHERE" is opened in Harajuku.	Reebok推出「INSTA PUMP FURY」 Reebok releases the INSTA PUMP FURY.
	NIKE推出「AIR HUARACHE」 Nike releases the AIR HUARACHE.		Michael Jordan宣布自NBA退休。轉而進軍棒球界 Michael Jordan announces his retirement from the NBA, and becomes a baseball player.	

Reebok_Insta Pump Fury

NIKE_AIR FOOTSCAPE

PRO-Keds_Royal Plus

asics_GELMAI

1990's

1999

1990年代中期～後期 裏原宿風格成為主流
Mid to Late 1990s: Harajuku's Underground Scene

1990年代後期 流行高科技球鞋
Late 1990s High-tech Sneakers

1995

NIKE推出「AIR MAX 95」
飆漲的市場價格及「狩獵AIR MAX」等
發展成一連串社會現象

Nike releases the AIR MAX 95. Prices escalate and cases of robbery of become a social issue.

NIKE推出「AIR FOOTSCAPE」

Nike releases the AIR FOOTSCAPE.

藤原浩開始於雜誌《Men's Non-No》連載「A Little Knowledge」專欄

Hiroshi Fujiwara starts the series featured in Men's Non-No magazine, named "A Little Knowledge".

1996

PRO-Ked's的哥倫比亞工廠關廠。
眾多球鞋迷開始搜尋「最後的哥倫比亞」

Keds Colombia plant closes down.
Fans go on a hunt for "Last Colombia".

「CHAPTER」球鞋店於原宿開幕

The store "CHAPTER" opens in Harajuku.

NIKE和Tiger Woods簽約

Nike signs Tiger Woods.

1997

「READY MADE」於原宿開幕（1999年歇業）

The store "Ready Made" opens in Harajuku. (Closed in 1999)

野茂英雄活躍於大聯盟。
NIKE首度推出日本人簽名鞋款「NOMOMAX」

Nomo Hideo makes his mark in the major league. The first Japanese signature model, the NOMO MAX is released from Nike.

1999

NIKE「DUNK」推出令人期待的復刻版。
將配色顛倒設計「裏DUNK」成為一時的話題

Long-awaited re-release of the Nike DUNK. The URA DUNK's reversed coloring attracts many people's attention.

NIKE推出「AIR KUKINI」

Nike releases the AIR KUKINI.

Gravis誕生。
和GOODENOUGH的聯名合作等造成不少話題

Gravis debuts.
The collaboration with GOODENOUGH attracts many people's attention.

Gravis_Rival

NIKE_AIR MAX 98

NORTH WAVE_Espresso

UNDERCOVER

藤原浩在球鞋界以指標人物散發出的強大影響力，就連球鞋品牌知名廠商NIKE也注意到了。藤原浩的立場從「接收」
轉變為「創造」層面，以HTM的系列開始陸續提攜案多經典鞋款的創作。

Hiroshi Fujiwara's strong influence as a key opinion leader in the sneaker industry was enough even to draw Nike's attention.
Hiroshi Fujiwara changed his position from "Consumer" to "Producer" and became involved in producing numerous classic sneakers, including HTM.

2000

2000年代前期　國產球鞋品牌的蓬勃

Early 2000s: The Rise of Domestic Sneaker Brands (in Japan)

2000	2001	2002	2004
NIKE推出「AIR PRESTO」 Nike releases the AIR PRESTO.	中村HIROKI開始經營 visvim品牌 Hiroki Nakamura launches visvim.	藤原浩於『Men's Non-No』的連載專欄「A Little Knowledge」首度發表HTM系列鞋款。 Hiroshi Fujiwara introduces HTM's first collection in a series featured in Men's Non-No magazine, "A Little Knowledge".	NIKE推出由藤原浩提案的配色款「ORCA PACK」。 Nike releases the ORCA PACK designed by Hiroshi Fujiwara.
NIKE推出「AIR WOVEN」 Nike releases the AIR WOVEN.	今井崇開始經營 MADFOOT!品牌 Takashi Imai launches MADFOOT!.		
NIKE發表技術革新的彈性系列「SHOX」 Nike announces their revolutionary cushioning system called SHOX.	NIKE推出由藤原浩所提案設計的配色鞋款「monotone collection」 Nike releases the Monotone Collection designed by Hiroshi Fujiwara.		《SHOES MASTER》創刊 First issue of "SHOES MASTER" is published.
「atmos」於原宿開幕 The shop "atmos" opens in Harajuku.			因新潟縣中越沖發生的地震災害，NIKE和藤原浩共同發起慈善拍賣募款 Nike and Hiroshi Fujiwara organize a charity auction to raise funds for victims of an earthquake at Nakakoshi offshore of Niigata.

NIKE_AIR PRESTO

NIKE_AIR WOVEN

NIKE_AIR ZOOM SEISMIC
[monotone collection]

NIKE_HTM AIR FORCE 1

2000's

2000年代中期～後期　Fast Fashion的興盛
Mid to Late 2000s: The Prosperity of Fast Fashion

2000年代後期　流行女性球鞋
Late 2000's: Ladies Sneakers

2005

以藤原浩、清永浩文、中村HIROKI 3人為中心，開始運作Web雜誌《honeyee.com》
Hiroshi Fujiwara, Hirofumi Kiyonaga and Hiroki Nakamura launch the Web Magazine "honeyee.com".

2007

fragment design和CONVERSE共同合作推出「ALL STAR」
Collaboration model between fragment design and Converse, ALL STAR is released.

《honeyee.mag》創刊。
First issue of magazine, "honeyee.mag" is published.

NIKE迎接「AIR FORCE 1」25周年，於原宿開幕了期間限定店舖「1 LOVE」，也在世界各主要城市舉辦紀念活動。
The Nike AIR FORCE 1 marks its 25th anniversary. Anniversary events are held in major cities around the world including the opening of the shop "1LOVE" in Harajuku for a limited period.

2008

藤原浩與SOPH.的共同品牌uniform experiment開始運作
Hiroshi Fujiwara and SOPH join forces to launch a new brand, UNIFORM EXPERIMENT.

因中國大陸四川省發生大地震災害，NIKE和藤原浩共同舉辦慈善拍賣募款
Nike and Hiroshi Fujiwara hold a charity auction in the aftermath of the earthquake in Sichuan, China.

藤原浩和村上隆共同舉辦《Hi & Lo》展。此活動的紀念商品為visvim推出的球鞋
Hiroshi Fujiwara and Takashi Murakami hold the exhibition, "Hi & Lo". Anniversary shoes are released from visvim in commemoration.

「NSW STORE」於原宿Cat Street開幕（2010年遷址）
pop-up store "NSW STORE" is opened on Cat Street in Harajuku. (2010 NSW comes out with another pop-up store at HP annex as "NSW at HP+")

2009

為了紀念《SHOES MASTER》創刊5周年，發行第一本書籍《Sneaker Tokyo》。
First volume of "Sneaker Tokyo" is published to mark the 5th anniversary of "SHOES MASTER".

「NIKE FLAGSHIPSTORE HARAJUKU」於原宿車站前開幕。
NIKE opens flagship store in front of Harajuku Station.

visvim_FBT

CONVERSE×fragment design_ALL STAR

NIKE_ORCA PACK

NIKE×fragment design_TENNIS CLASSIC

Long Interview with Hiroshi Fujiwara

藤原浩　長篇專訪

作為本書的總結，我們大膽進行了藤原浩的個人超長篇專訪。
從第一次穿的球鞋、記憶最深刻的球鞋，
還有今後的潮流預測等，只要與球鞋相關的種種話題，我們都追根究底問清楚。

To conclude this book, a long interview with Hiroshi Fujiwara himself.
We hear from Hiroshi on many things about sneakers, from the first pair he wore, to his most memorable pair,
and his thoughts on sneaker trends to come.

採訪者：榎本一生（《SHOES MASTER》總編輯）
Interviewer: Issey Enomoto (Chief Editor of SHOES MASTER)

——在這本書中，我們介紹了藤原先生至今穿過的球鞋及創作過的球鞋款式等，而在這最後的總結，我們想詢問藤原先生的是關於球鞋的一切。首先，藤原先生是否可以告訴我們你穿的第一雙球鞋是什麼？

「之前我也在部落格中寫過，小學運動會的時候我穿的是足袋，這是真的哦。關於這個也可以去問DETZ。」

——嗯，先將這個話題擺一邊（笑）。

「有個叫做PANTHER的品牌在當時很流行，我也穿過，那應該是我的第一雙球鞋吧。款式設計和Onitsuka Tiger很相似，顏色是水藍色。」

——欸～有過這種球鞋啊？

「嗯，不知道該說是球鞋還是運動鞋。」

——藤原先生是哪一年出生的啊？

「1964年。」

——沒錯的話《POPEYE》創刊是在1976年，那時候您是中學生吧？

「是的。當時我也買了創刊號。」

——那個世代的年輕人深受《POPEYE》影響，藤原先生之所以喜愛球鞋也是因為如此嗎？

「嗯，如何回答呢……看你對於『球鞋』的定義是甚麼。在《POPEYE》發行前，例如VANS的鞋款我也會因為喜歡而經常穿著，就如被稱為『356』Sagoroku的款式。」

——那麼這是不是你的第一雙我們今天稱之為「球鞋」的鞋款呢？

「很難說呢，我第一雙球鞋有點像CONVERSE。還有，REGAL在IVY風格盛行時也推出這類似TOP-SIDER帆船鞋的球鞋，我也穿過。」

——IVY風格盛行的時候，藤原先生是中學生嗎？

「是小學高年級到中學低年級。」

——Throughout this book, we have introduced many of the sneakers you have worn and designed, and now to finish, I'd like to ask you about all things connected to sneakers. First of all, can you tell us about your first pair of sneakers you wore?

"I wrote about this in my blog before, but for sport's day at elementary school, I wore tabi. (Japanese traditional style socks, often wore with kimono.) Honestly!"

——OK… (laugh) apart from tabi…

"When I was a kid, I used to wear shoes made by a brand called Panther. They were popular at the time. I think those were my first pair. They were similar to Onitsuka Tiger. In a light blue color."

——Really? I don't think I've heard of them.

"Well, they were more like gym shoes than actual sneakers."

——What year were you born in?

"1964"

——If I remember correctly, Popeye was first published in 1976. You must have been a junior high school student then?

"Yeah, I bought their first issue."

——I would imagine that quite many people of your generation were heavily influenced by Popeye magazine. Would you say your interest in sneakers originated from Popeye in some way?

"Mmmm… Depends on what you mean by originate I guess. Although even before Popeye, brands like Vans were around. I used to wear a model by them called the 356. Everyone used to call them "Sagoroku"." (Japanese abbreviation of three-five-six)

——Were they the first pair of shoes you owned we would today call sneakers, as opposed to gym shoes?

"It's hard to say. I guess they were similar in design to Converse. Also Regal was coming out with sneakers like the Topsider. They were a type of deck shoe. They were popular during the "Ivy" trend. (Ivy refers to a period in Japan when the fashion style of US college students was popular.) I used to wear them myself."

——Were you at junior high at the time of the Ivy boom?

"It was probably around my later years at elementary school to early years

——小學生就跟隨IVY風潮，真的是很早熟的小學生啊。

「那個時候，關於流行和音樂等東西都是受到我姊姊的影響。」

——中學時有參加什麼社團活動嗎？

「我參加籃球社。當時鞋款都以ONITSUKA為主。1年級都穿帆布，升上2、3年級後可以升級穿麂皮材質的FABRE。FABRE有很多配色，可用來劃分隊伍顏色，另外有些時髦的人會穿著CONVERSE或adidas。」

——藤原先生當時是穿什麼呢？

「我穿的是灰色麂皮CONVERSE ONE STAR，另外也穿過PRO-Keds ROYAL PLUS，這雙也是灰色的麂皮材質。」

——在當時ONITSUKA成為主流的情況下，選擇CONVERSE或PRO-Keds的人應該比較少吧？當時是國外品牌球鞋本身就很稀少的時代。

「嗯，或許是。當時NIKE還不是主流，我是看了《POPEYE》刊載的『美國屋鞋店』廣告，再郵購NIKE球鞋。再來是玩滑板時常穿的VANS，當時幾乎沒有滑板專用鞋，開始時穿的是紅和深藍雙色款，之後也穿過亮色系的鞋款。」

——VANS是在哪裡購買的呢？

「大阪有一間叫做『SPORT TAKAHASHI』

at junior high."

——You must have been a fairly sophisticated elementary school kid to be into Ivy.

"My taste in fashion and music at that age was heavily influenced by my older sister."

——Where you a part of any clubs when you were at junior high?

"I was part of the basketball club. At that time, Onitsuka were the most popular shoe brand. In the first year everyone was wearing them in canvas, then in second and third year, everyone stepped up and switched to suede. There was a model called Fabre. It came in a range of different colors so people used to get them in their own team color. Then, there were more fashionable kids who wore Converse or Adidas."

——What were you wearing?

"One Star from Converse in gray suede. Also the Royal Plus by PRO-Keds. They were also gray suede."

——There must not have been many people wearing Converse and PRO-Keds if Onitsuka was the main trend at the time? Overseas brands must have been pretty rare.

"That's perhaps true. Nike wasn't a mainstream brand at all back then. I found an ad for an American shoe store in an issue of Popeye and used to buy Nike mail order. I also used to wear Vans a lot for skateboarding. It was still an era when there were hardly any skateboarding shoes around. At first I was wearing Two Tone red and navy, then later I also wore them in Crazy Color."

中學時代參加籃球社，穿的是麂皮材質的 ONE STAR。

During his years at junior high, Hiroshi played basketball in suede One Star.

——當時周遭的人都穿什麼球鞋呢？

「籃球社的人當然堅持穿籃球鞋，而玩滑板的人也就堅持穿滑板鞋，不過這都是因為機能而做出的選擇。我雖然對於機能也十分講究，不過外形對我來說也是同等重要的。我認為都是因為《POPEYE》的強大影響力。」

——這是「西海岸文化」進入日本的時期吧，進入高中後又如何呢？

「還是VANS佔多數。那時候不僅是大阪，我偶爾還會到東京買鞋。」

——也就是受到RUN D.M.C.的影響而非常流行adidas的那個時期嗎？

「雖然RUN D.M.C.的影響很大，不過在這之前已經有HIP HOP的流行風潮，我也模仿國外的HIP HOP藝人將PUMA換上粗鞋帶，或是穿著PATRICK的尼龍黏帶鞋款。我記得Malcolm McLaren也穿過。」

——來到東京之後，都去哪裡尋找球鞋呢？

「東京的話還是阿美橫吧，還有就是逛逛不同地方的體育用品店吧。」

——以前在尋找球鞋時會有一種尋寶的樂趣呢。

「是啊。以品牌來說，當時還是以adidas最有人氣。NIKE的話則是和Jordan合作之後才流行吧！」

——AIR JORDAN的第1代上市是在1986年。

「當時LL COOL J穿過，然後玩滑板的人也穿AIR JORDAN了。」

——藤原先生當時穿NIKE球鞋玩滑板，挑選時有哪些重點要注意呢？

「鞋底啦。重點是鞋底要薄，所以像AIR FORCE 1就不太適合玩滑板。AIR JORDAN的第1代鞋底則偏薄，非常適合滑板運動。不過最好的鞋款

——1992年巴塞隆納奧運時，JORDAN率領NBA超級球星組成「夢幻球隊」，感覺上球鞋的大流行從那時開始加速。

「我認為球鞋的流行是在1980年代的HIP HOP風靡時開始，後來才延續至滑板和籃球運動。」

——球鞋流行經歷過好幾次風浪。從1980年代的HIP HOP衍生出adidas和PUMA的流行，而1990年前期則有和NBA連結的籃球鞋風潮。

「如果說到時尚潮流，當時還有DC品牌的流行。」

——DC品牌，流行過呢。那個時候藤原先生是穿著什麼服裝呢？

「因為姊姊的影響，我也穿過BIGI等品牌，之後有一些川久保玲或山本耀司的服飾，並沒有完全沉醉於DC中。我喜歡的是衝浪風格或是龐克風格的造型吧。」

——那麼選擇什麼款式的鞋子呢？

「我也穿球鞋，也常穿在ROBOT買的厚底鞋，還有Dr. Martens等。」

——接著是1990年代以AIR JORDAN、AIR MAX等為代表的高科技鞋款的流行時期。

「以那個時候為界線，我覺得球鞋設計有了革命性的改變。無論是外形或機能方面。」

——陸續推出的是帶有未來感設計的鞋款。

「對啊。雖然那個時候的款式設計很新鮮，但我個人是不怎麼喜歡的。」

——材質方面，原本只有皮革或帆布等材質，也開始變成使用翻毛皮革或人工皮革等，或是AIR JORDAN也採用了塑膠材質的零件。

「There was a store in Osaka called Sports Takahashi. I used to go by train from my hometown, Ise."

——What shoes were the people around you wearing?

"People into basketball were wearing basketball shoes and people into skateboarding were into skateboarding shoes, but purely for functional purposes. I also liked sneakers for how they performed as a sports shoe, but the look of the shoes were of equal importance to me. No doubt Popeye had a great influence on me in that sense."

——I believe it was around the time West-Side culture started coming into Japan. What about after you started at senior high?

"I guess I wore Vans the most. I started taking shopping trips up to Tokyo around that age."

——Was that also around the time Adidas became popular as a result of the influence of RUN D.M.C.?

"RUN D.M.C.'s influence was also huge, but the hip hop boom had already started when they came on the scene. I used to imitate overseas hiphop artists by wearing Puma's with fat shoelaces, or Velcro Patricks. I think Malcolm McLaren was wearing Patricks at the time."

——What types of stores were you buying sneakers from since you made the move to Tokyo?

"Ameyoko I guess. I used to look round the rural sports outlets too"

——In the old days shopping for sneakers definitely had a treasure hunt sort of appeal to it.

"Yeah. Adidas was by far the most popular brand back then. Nike didn't really come on the scene till Michael Jordan I guess."

——The first model for Air Jordan went on sale in 1986.

"LL Cool J was wearing them at the time. Skaters started wearing them for skating around the same time."

——I understand you also wore Nike for skateboarding. Can you tell us why exactly you chose Nike over other brands?

"Nike shoes have a thin sole. Shoes with a chunkier sole like the Air Force 1 are not so good for skating. The thinner sole of Air Jordan's were very much suited to skate boarding. But perhaps the best shoes were Vans, for their soft sole. In those days tricks like the ollie were yet to be conceived, so there was no risk of the upper tearing. The sole was the only part of the shoe that got any abuse."

——In the 1992 Barcelona Olympics, a lot of NBA stars like Jordan arrived on the scene as the Dream Team. I think it was around that time that the sneaker boom really began to pick up pace.

"For us the sneaker boom really started before basketball, from around 1980 when hip hop became popular. Skating, and then basketball were kind of an extension from that for me."

——There were a few phases to the sneaker boom. Hip-hop from the 80s made brands like Adidas and Puma popular and then in the early 90s there was a basketball shoe boom.

"In the fashion world DC were popular for a while too, weren't they."

——The DC brand was extremely popular for a while. What kind of clothes were you wearing at the time?

"Influenced by my sister, there was a time I wore BIGI. After that I also owned a few things from Comme des Garcons and Yohji Yamamoto, but never really got into DC. There were a few styles going around at the time. I was more into surfer and punk style clothing."

——What were you wearing on your feet?

"I was wearing a lot of sneakers, but I also often wore a pair of rubber soles I bought from Robot. Also Dr. Martens."

——Then in the 90s there were Air Jordan's and the Air Max, two exemplary shoes from the hi-tech sneaker boom.

"I think that was a revolutionary time for sneakers, both technology wise and appearance wise."

——There were a lot of cutting edge designs coming out at that time.

"Yeah, although personally I wasn't really into those designs myself."

——Materials wise, until then sneakers were mostly either made from leather or canvas, but then nubuck and artificial leather materials appeared, then in the case of shoes like Air Jordan, plastics also started to be used as well.

"Yeah, that sums up basically what was happening in that era."

——另外，藤原先生當時經常在雜誌介紹自己喜歡的球鞋，因此被稱為「球鞋界的潮流領導者」、「藤原浩介紹的球鞋絕對會流行」等，聽到這類的聲音，你有何感想呢？

「嗯……這不僅球鞋，我每次都是這麼說的，不是因為我所看過、穿過的會流行，而是對於將會流行的事物，我都比一般人更早有反應，就像NIKE成功找了擁有這種察覺能力的我，把新商品早一步送到我手上，目的是為了便捷的宣傳效果。」

——原來如此。

「但是，這不限於球鞋，經常都有企業廠商來拜託我『請試用看看』、

『請在雜誌中介紹』，不過如果不是我真正喜歡的東西，我是不會介紹的，素來我都是如此，所以我的說服力或許是因為這樣產生出來吧。」

——我認為這是很大的原因。在honeyee.com經營部落格後，來自企業廠商的邀約應該增加不少吧？

「嗯。這是我現在停止部落格的原因。部落格漸漸變得商業化，總是會變成『請幫我們刊登在部落格上吧』的情況，我就不太喜歡了。」

——honeyee.com的「PLOG」也是為了因應這種情況而產生的嗎？

「也有這種考量，不過也是因為在

——You have been introducing your favorite sneakers in magazines since then. In the industry it's often said that if trend reader Hiroshi Fujiwara introduces a pair of sneakers, they are sure to become a hit. What do you yourself think of this influence you seem to have?
"It's not just sneakers I've introduced, but I always say, it's not that the stuff I introduce becomes popular because I've seen or worn them. I think I just pick up on trends that little bit earlier than other people. It's strange for me to say this but I think Nike have used me well in that way. I think they are aiming for the subliminal promotion they know I can provide by sending me their new products at an early stage."
—— I can understand that.
"I get a lot of companies wanting to use their products or introduce their products in magazines, but I don't like to introduce a product if I don't actually like it, so I've never done so. I think that's why people can trust my opinion."
——I agree that's a huge factor. Have such offers from companies increased since you started your blog on honeyee.com?
"Yeah, that's one of the main reasons I stopped writing on my blog. The blog became too commercial. I was being asked to write about products on my blog all the time. It put me right off."
——Was the "PLOG" section on honeyee.com meant for such purposes?
"That's part of the purpose of the PLOG, but the products being introduced

對於將會流行的事物，
我都比一般人更早有反應。

Hiroshi, always a step ahead to pick up on the next big thing.

『PLOG』中介紹和由我本人親自介紹，意義上還是不同。如果是知名部落客，也就是背負著某種招牌在經營的人，我很能理解他們會變得商業化。因此，我現在回歸到平面媒體的部分，我會在《SENSE》連載專欄介紹我注意的東西。雖說如此，和網路相比平面媒體傳播的速度沒有那麼快，所以在看到我的介紹之前可能在其他地方已經被頻繁介紹過了。」

——原來如此。不過，現在重新看以前的報導，我覺得藤原先生對於球鞋的喜好似乎並沒有太大的改變。

「或許是吧。喜好是不會有太大的改變的。偶爾會因為新的東西產生新奇感然後變心一下而已。」

——觀看這次本文中介紹的球鞋，感覺十分統一、完整，就好像它們之間有著什麼共通點。雖然我無法更具體的知道，但以藤原先生本身來說，有喜歡或不喜歡的類別嗎？

「或許我還是喜歡籃球鞋吧。如果要說不太喜歡的鞋款，我對於慢跑鞋比較沒有什麼反應，我連AIR MAX 95都沒穿過。」

——的確內容中很少出現慢跑鞋。

「不過我滿喜歡KUKINI或SEISMIC，還有不知道該說它是慢跑鞋嗎，PRESTO給我很大的衝擊。」

——PRESTO給人的印象很深刻啊。

「當時的感想是居然會有這種球鞋啊。雖然是高科技鞋款，但是外形不像高科技這點還滿好的。還有我也滿喜歡FOOTSCAPE的，對了，這是什麼時候上市的？」

——第一次上市是在1995年。

「和AIR MAX 95同年啊。這樣說來，當大家都在注意AIR MAX 95的時候，我常穿的卻是FOOTSCAPE啊！」

——但FOOTSCAPE也不能說是慢跑鞋，是很難分類的鞋款啊。

「或許是哦。說到這個，當時有些配色是只有女生鞋款才有的，我就會去找女生款的大尺碼。」

——AIR MAX 95也是如此。不少女生款才出現的配色都很好看呢。

「對，就像我記得FOOTSCAPE的紫色就只有女裝呢。」

on the PLOG are not products that I myself am introducing. There are a lot of people who write blogs for their own personal purposes, but there are other people who write with the intention of promoting something. These blogs normally end up going commercial, which can't be helped. That's one of the reasons I've gone back to paper, now introducing products that catch my eye in serial publications like SENSE. Having said that, compared to the web, printed publications take more time to get out, so often a product that I'll introduce in a magazine or whatever will be all over the web before the article's published"
——Looking back on articles you featured in the past, it doesn't seem like your taste in sneakers has changed so much over the years.
"I don't think a person's taste changes much. Apart from the odd fling when something new and interesting appears."
——Just looking through the shoes that we have featured in this book, it feels like there is a common theme or some combining element to the shoes you select, although I'm unable to put my finger on it. Are there certain sneakers you like or dislike?
"I would say I like basketball shoes. I was never a big fan of running shoes. I never wore the Air Max 95 (although they were popular)."
——You're right, we hardly featured any running shoes in this book.
"Although I did like the Kukini and Seismic. Also, I'm not sure you would call them running shoes but the Air Presto had a big impact on me as well."
——Presto was a strong shoe.
"I'd never seen a shoe like it before. It was a high-performance shoe, but at the same time it didn't appear to be, which was something I liked. I liked the Footscape too. When did that come out again?"
——Initially in 1995.
"The same year as Air Max. I guess that means when everyone was going mad about the Air Max, I was wearing the Footscape."
——I wouldn't call the Footscape a running shoe, but it's a difficult sneaker to categorize.
"Yeah. They were making a lot of colors only available in women's, so I remember I went searching for them in large sizes."
——The Air Max was the same. There were a lot of good colours but they were only available in women's.
"Yeah, I remember the Footscape in purple was only available in womens."

——轉換一下話題，本次這本書中揭載了藤原先生穿過或創作過的球鞋共約200雙，幾乎全部都是藤原先生的私人收藏。而為了拍照而將所有球鞋排開時，老實說真的嚇了我一跳，為什麼可以將以前的東西保存得如此完整？

「不，與其說保存，不如說找不到適當時機丟掉吧。」

——而且，還能以品牌區分並且完整的保存，這真是太厲害了。

「不僅是球鞋，基本上我是不會丟東西的人。如果不丟掉並將它保存好，或許有一天又會出現我自己想穿某一件衣服或鞋子的時候。」

——雖說如此，但是我感覺藤原先生和「收藏家」不同。收藏家通常會以欣賞為樂趣，有些人還會擺放出來。這些人之中，有些收藏家還會覺得很珍貴的球鞋或牛仔褲拿出來穿太浪費。但是，藤原先生就不一樣了，喜歡的衣服和球鞋就盡量穿。

「與其自己一個人擁有欣賞的樂趣，不如分享出來，或許我是這樣子的人。」

——順帶問一下，您能掌握自己擁有多少雙球鞋嗎？

「完全不能掌握。對了，復古風潮盛行的時期，球鞋的價格飆漲得很厲害，我曾經賣掉過幾雙鞋。」

——We've featured about 200 pairs of the shoes you have worn and designed, and the great majority of them are shoes you actually own. When I went round to pick them up to photograph, I was honestly quite shocked. Why do you keep holding on to shoes you wore such a long time ago?

"It's not really intentional. It's just I never found the time to throw them out."

——It's amazing how you store them organized by brand.

"I'm not the type to throw sneakers or anything away. I'm always aware that if I keep holding on to something, the time I want to wear it might come round again."

——Having said that, I think you're different from the average collector. There are a lot of collectors who display their shoes just so they can enjoy looking at them. Some collectors think it's a waste to wear the sneakers or denim that they like. But you don't seem to have any problem wearing items till they are worn out.

"I would say I'm more the type who prefers to share things with other people, rather than simply enjoy looking at the things I own."

——Do you know how many pairs you have in total?

"No idea. A while ago there was a vintage boom and sneakers were selling for ridiculous prices. I sold a few pairs at that time."

——Really? What and how did you sell them?

"I sold a couple of pairs of Air Jordan's first line, one in blue and black and

我不會丟掉球鞋，因為或許有一天會再次出現自己想穿的時候。

Don't throw shoes out,
as the day you want to wear them again may come round again.

——啊？賣掉什麼？在哪裡賣的呢？

「是AIR JORDAN第1代的藍×黑和灰×黑的滯銷品，在以前的『LONDIS』賣掉呢。」

——這地方真是令人懷念，是以「藤原浩的私有物品」販賣嗎？

「不是這樣的。是和其他的商品一起銷售，我記得賣的價錢很好哦。」

——那麼，現在擁有的球鞋以後有意願讓出來嗎？例如，如果有品牌廠商希望當作資料保存的話，您會怎麼做呢？

「如果是這樣的話，對於以後應該不會再穿的球鞋，讓出來也可以。」

——在這本書的序提過，內文揭載的藤原先生的球鞋檔案，有非常高的資料價值，但願品牌相關人士會看到，然後重新發現這些球鞋的好地方，把它們重新發行就好了。

「嗯，在我個人的想法中，這本書中出現的鞋款，其實有很多已經是常被重新發行的商品，要看通不同世代的想法是不容易的事情，例如我看到70年代的網球鞋，心想如果能把它

再推出就好了，但你要知道20多歲的年輕人是否有相同的想法是一件困難的事。某種意義上，HTM扮演的就是這種角色，先以HTM的名義作為實驗性的復刻，然後再而變成Nike自家的商品。」

——但是現在想要HTM WOVEN BOOTS也買不到啊。

「啊，對啊，像這種商品如果能再製造推出就好了。」

——如果要將現有的球鞋依照喜歡的順序排列，最喜歡的是哪一雙呢？

「要排序真的很難，不過我最喜歡的是DUNK或COURT FORCE吧。」

——那應該也包含了以前玩滑板的回憶吧。

「啊，是的。」

——相反，有不喜歡的球鞋款式嗎？例如，我沒有看過藤原先生穿鞋底也是黑色的全黑球鞋。

「啊，或許是的，我應該沒有穿過。AIR JORDAN 5和6也是鞋底全部都是黑色的款式吧，我就不是很喜歡。」

another in grey and black, at Londis. The store's not around anymore though.
——Londis, that takes me back. Did you sell them as your own personal possessions?
"No, they were sold as part of a collection with some other items. I think they got a good price though."
——What about the shoes you own now? Do you intend to keep a hold of them? What would you do for example if a manufacturer came to you and said they wanted to have a pair for their own archive?
"If I was approached and I didn't have any intention of wearing the shoes again, I would probably give them up."
——I made the same comment in the prologue of this book but, I would imagine the shoes featured in the archive section would fetch quite a hefty price. It would be nice if somebody working for one of the brands was reminded of the quality of any of the shoes featured and was convinced to re-release them.
"I think a lot of the shoes in this book are constantly being re-released already. It's difficult to read the generations. For example when I look at a pair of tennis shoes that were around in the 70s and feel they should be re-released, it's hard to know if a person in their 20s would feel the same way. HTM in a sense has that role. Shoes that are revisited are first released as an experiment under the HTM name, before they are released as an inline Nike product."
——But at the same time, some products like HTM's Woven Boots are no longer available to buy.
"Yeah… It would be good if we could do that model again."
——If you were to line up all the shoes you had in order of preference, which would be your favorite?
"It would be difficult to put them in order, but my favorites are Dunk and Court Force."
——Is that because you have a sentimental attachment to them from when you wore them for skating?
"I guess so."
——Are there any types of sneakers you simply don't like? For example, I don't think I've ever seen you wear a pair of all black sneakers from the sole up.
"No, I don't think I have. Air Jordan's 5 and 6 were all black, weren't they? Wasn't keen on those."

——即使如此，今後會出現什麼樣子的球鞋呢。

「現在是以偏向網球鞋樣式的鞋款為潮流，還有是VANS或CONVERSE這類，換句話說是不強調高科技，至於接下來就要看這股潮流如何轉變。」

——如果還能產生有趣的流行風潮就好了，不過我覺得不會再出現像AIR JORDAN或AIR MAX的大浪潮了。

「最近只要有什麼引起潮流，馬上就有企業仿效，如此一來真正的本質就會變得淡薄，我認為這就是困難的地方。我自己竭盡所能地將『真正自己想穿的款式』提案給NIKE。最近推出的ALL COURT CANVAS也是我認為如果能有這種款式的球鞋應該不錯才推出的。ALL COURT最早是由A.P.C.的Jean推出復刻版，我看了後也覺得鞋尖的設計很不錯呢。」

——藤原先生去年也向NIKE提案了類似鏡面皮革質感的鞋款，接下來或許可以利用在CANVAS鞋款上，應該不錯吧。

「CANVAS也有多種不同的質感，我個人比較想利用的是細紋路而不是粗紋路的材質。」

——沒錯，這款**ALL COURT**的**CANVAS**的確紋路很細緻。

「還有，這雙CANVAS ALL COURT的設計重點是沒有SWOOSH。關於這點剛開始時還被法務部門的人說絕對不可以沒有SWOOSH，但是，實際又如何呢？我問了Fraser和Atsuyo的意見，也直接跟Mark提出我的想法，最後終於實現了。」

——說不能沒有SWOOSH的**NIKE**高層或是**Mark Parker**，從他們的立場來看，或許會認為為什麼如此堅持不要呢，有或沒有都一樣吧。

「搞不好真的被這麼認為。剛開始做HTM AIR FORCE 1時也曾說過不要放SWOOSH，但結果還是不行。」

——現在再看沒有SWOOSH的**ALL COURT**，就覺得的確有沒有SWOOSH比較好看，但是對於不了解這種感覺的人來說，接收到這樣的提案或許真的很難理解。

「我跟Nike的合作及溝通已有一段很長的時間，所以現在他們應該非常清楚了吧，像我喜歡的一系列白色鞋底的鞋款，包括Mark等的工作人員剛開始也是不能理解我的想法。」

——What do you see for the future of sneakers?

"At the moment we're seeing a lot of tennis shoe type sneakers. What we call low tech shoes such as Vans and Converse are also popular. It all depends on where that trend goes next."

——It would be good if we saw another interesting movement in sneakers. Although I don't sense we will see any massive movements like the ones that we have had in the past, like with Air Jordan or Air Max.

"Recently as soon as a new trend appears on the market, the manufacturers jump onto it almost immediately. It means that we lose a lot of the genuine essence of the trend. It's a difficult area. I myself try as much as possible to promote sneakers that I genuinely want to wear. All Court Canvas shoes that we released recently is another example of a shoe that I saw and I liked and developed an idea for. Jean from A.P.C. brought the shoes back first, and I liked the design of the toe so..."

——You also proposed to Nike a model made in glass leather last year, and this year you made the suggestion they bring out a canvas version of the same design.

"There are many types of canvas. I wanted to make the shoes in a finely woven canvas, as opposed to a more coarse canvas."

——The canvas for these All Courts is certainly finely woven.

"Another point about this pair is that we made them without the Nike swoosh logo. At first I was told it simply wouldn't pass legal. But then when I asked Fraser and Atsuyo and spoke to Mark directly about it, we managed to do it in the end."

——This may sound trivial, but looking at it from the CEO of Nike, Mark Parker's perspective, didn't he wonder why you wanted to lose the swoosh logo? Didn't he think that that was a rather minor detail for you to be focusing on?

"He may have thought that. Because I have tried to take the swoosh out from the HTM Air Force 1 once but it never got approved."

——Looking at your new All Court without the swoosh logo, I definitely think they are cool. However, I also think that this execution may be difficult to understand to some people.

"I think I've been with Nike long enough to share our idea very well. You can see that from series of white outsole shoes we work together."

——The color of the sole of a shoe can make a big difference so I can understand that, but whether a shoe bears the swoosh logo or not...

——鞋底是白色或黑色有很大的差別，這點我能理解，可是有沒有SWOOSH是非常微妙的差別吧？

「是嗎？我不覺得很微妙。因為很明顯沒有SWOOSH比較好看吧？」

——嗯，像你可以感覺到「很明顯的沒有SWOOSH比較好看」這點，是有點東京的特質，或是可以說是帶有藤原先生的特質吧，我覺得這不是每一個人都能明白的。換句話說，可以將此稱為「減法的美學」吧。

「從開始就讓人看見品牌的力量而成功的商品很多，但是我追求的不是這個，例如有人看到後會問這雙鞋很好看耶，是什麼牌子的呢？然後再告訴他其實是NIKE哦，我認為這種不刻意的感覺才能將優點表現出來。」

——球鞋中有很多是特別版本都會極力強調「這是我們創作的」，但藤原先生過去所設計的卻完全相反，例如不需要放這個，不要這個比較好，都是類似以減少設計要素的方向製作商品，關於這點就有很大的不同吧。

「在我對於美的意識中，以真人不露相的方式表現比較有型，或許這點就非常像日本人吧。」

——是像日本的幽雅感吧。這麼說的話，在我開始注意球鞋的1990年前後當然沒有HTM，也沒有日本球鞋店舖或品牌的別注版本。然後，將此市場拓展開來的是海外球鞋店舖的別注鞋款，而且很明顯知道這是「Foot Locker特別訂製款」。

「我也常常買Foot Locker特別版。因為有很多配色是只有它才有。」

——但是這就不是「減法美學」，而是「加法美學」設計吧。

「那時應該還沒有想得這麼深奧吧？（笑）因為Foot Locker是很大的客戶，所以製造的都是熱賣款式，也不會強調是誰設計的呢。」

——回想以前的Foot Locker特別版真的令人十分懷念。不過當我第一次去美國的店時卻有一點失望。（笑）

「美國大城市如紐約等，大部分都有Foot Locker，每一家的陳列商品都是一樣，所以就算當時覺得這雙鞋不錯哦，又會想任何時候都買得到，下次再買就好了，但當你想買的時候卻又沒尺寸。雖然不僅限於球鞋，但買東西的時機很難拿捏啊。」

——雖然現在原宿周邊沒有Foot Locker，但是和以前相比卻增加了許多球鞋店，各大品牌的旗艦店都在此出現了，如NIKE、adidas、Puma、New Balance⋯⋯。

「還有許多小店呢。」

——還有ABC-MART等大型鞋店。藤原先生偶爾也會逛逛這種店嗎？

「我偶爾會逛，也會買哦。不久之前在ABC-MART購買了CONVERSE的ALL STAR，最近也在附近的ASBee買了鞋帶。」

——藤原先生也會去ABC-MART或ASBee買東西啊，沒有被店員發現你是誰嗎？

「完全沒有被發現哦，很平常的逛，很平常的買東西。」

——很叫人意外呢（笑）。最後想針對「藤原浩的球鞋」重新整理一下作一個總結，你喜歡的鞋款主要是藍球鞋或網球鞋⋯⋯。

「是的。」

——至於鞋身的配色最多是2色，鞋底為白色，鞋尖要帶有厚實感⋯⋯

「完全正確。這不是將我的底蘊全看穿了嗎！（笑）」

"To me it was quite significant. Looking at the shoe now, there's no doubt it's better without the logo right?"
——I think that sense for design is very Tokyo, or rather very Hiroshi. I don't think it's an opinion that can be so easily shared among others… that kind of "less is more" attitude to design.
"Some shoes sell because of the brand name attached to them. I believe the quality of a shoe can be better understood by people if they are drawn to a shoe before they find out it is made by Nike (or any other brand)."
——There are a lot of custom designed or collaboration sneakers out there, but most brands in the business are quick to stamp their name on products to promote the fact that the design belongs to them. But the shoes you have worked on in the past are the complete opposite. Your creations seem to come together through a process of removing elements you feel are unnecessary. It's something that's makes you very different as a designer.
"I think beauty should be simple and not give too much away. It's perhaps a very Japanese way of thinking."
——Your sense of beauty and grace is indeed very Japanese. I remember when I first woke up to sneakers around 1990, there was no HTM at that time of course and no Japanese shops or brands taking custom orders. Instead I was drawn to the custom order shoes I got from Foot Locker, a shop from the US. By looking at a pair of their custom orders, you can tell straight away that they're Foot Lockers.
"I used to buy Foot Locker custom orders often too. There were a lot of colours they only did for custom order."
——So Foot Locker is another good example. Like we were just saying, rather than "less is more", their motto seems to be "bold is better".
"I doubt they were thinking about it that hard. (laugh) Foot Locker are a big scale store, so they simply want to make shoes they know they can sell a lot of. There's no mention of who designed what with Foot Locker either."
——I was pretty infatuated with Foot Locker custom orders. I remember I was quite disappointed when I went to Foot Locker for the first time in the US though. (laugh)
"In New York and other big cities in the States, there are Foot Locker stores everywhere and they tend to hold mostly all the same stock. Even if you find a shoe that you like, you know you can buy it any time so you put off buying it for a while, until you return to the store and they are all out of your size. It's difficult to time it right. And I'm not only talking about sneakers."
——There's no Foot Locker store, but compared to the old days there are a lot more sneaker stores in and around Harajuku than there used to be. All brands seem to have their flag ship store there. Nike, Adidas, Puma, New Balance…
"A lot more smaller stores have opened up too."
——And of course there's mega stores like ABC Mart too. Do you ever drop by those kinds of sneaker stores?
"Sometimes I stop by. There are even times I will by the odd pair. Just recently I bought a pair of Converse All Stars from ABC Mart and the other day I some leather laces from ASBee."
——So Hiroshi Fujiwara shops at ABC Mart and ASBee as well? Did the shop assistant realise who you were?
"Not at all. I dropped by, did my shopping and was out of there just as normal."
——That's surprising. (laugh) To conclude this interview I would like to ask you one more time, looking back on the shoes featured in the Archive section of this publication, Hiroshi Fujiwara likes predominantly basketball shoes and tennis shoes…
"That's correct."
——You don't like more than two colours on one shoe, you like a white sole, and a sturdy toe bumper.
"That's all correct too. Stop stealing all my lines! (laugh)"

Abandoned Designs

藤原浩最初接觸NIKE時提案，但最後沒有實踐製作出來的早期HTM球鞋設計藍圖。
經過多年跟NIKE的合作，藤原浩先後把大量球鞋帶到世界的層面，不過不少設計還
是收藏起來等待重見天日的一天。

Some design ideas from the early stages of HTM that Hiroshi Fujiwara approached Nike with
that were put aside before they made it to the production stage. Through his collaborations
with Nike, Hiroshi Fujiwara has sent a great number of sneakers out into the world, but there
are also some designs that were shelved before they saw the light of day.

Index

References

本書撰文時，曾使用以下的書籍、MOOK、雜誌及網頁作為參考。

[書籍]
『スニーカー大図鑑』右近亨編／グリーンアロー出版 (1997)
『SOLE PROVIDER』Words by Robert "Scoop" jackson／powerHOUSE Books (2002)
『THE SHADOW OF THE OFFICIAL ART WORKS』HIROSHI FUJIWARA／FRAGMENT CO.,LTD. (2003)
『Blue RIBBONS』芥川貴之志著／河出書房新社 (2005)
『BRANDS A TO Z: adidas』Peng Yangjun, Chen Jiaojiao／ビー・エヌ・エフ新社 (2007)
『丘の上のパンク』川勝正幸編著 藤原ヒロシ監修／小学館 (2009)
『パーソナル・エフェクツ』藤原ヒロシ著／マガジンハウス (2009)

[MOOK]
『アディダス・ブック』ワールドフォトプレス (1998)
『永久定番 コンバース』祥伝社 (1999)
『NIKE 2001』辰巳出版 (2000)
『アサヤン増刊号 FootStep』ぶんか社 (2002)
『smart スニーカーBOOK』宝島社 (2003)
『KICKS STYLE vol.1-vol.3』祥伝社 (2001-2003)
『NIKE BIBLE 2006』宝島社 (2006)
『HUGE NIKE AIR FORCE 1 SPECIAL ISSUE / CODE+813 tokyo insight』講談社 (2007)
『スニーカージャックプレミアム ナイキ エア フォース 1』(2007)
『スニーカージャックプレミアム まるごと1冊エアジョーダン23周年』KKベストセラーズ (2008)

[雑誌]
『Men's Non-no』集英社
『Asayan』ぶんか社
『Hot-dog Press』講談社
『EYESCREAM』USEN

[網頁]
NIKE nike.jp
visvim www.visvim.tv
adidas www.adidas.com/jp
PUMA www.puma.jp
new balance www.newbalance.co.jp
CONVERSE www.converse.co.jp
Gravis www.gravisfootwear.com
VANS www.vansjapan.com
PRO-Keds www.prokeds.com
asics www.asics.co.jp
Reebok www.reebokjapan.com
mita sneakers www.mita-sneakers.co.jp
SKIT www.k-skit.com

特別在此感謝藤原浩先生，由本書籌劃內容、拍攝到撰文期間的協力與幫助。

Note from the Editor

我的初次球鞋回憶

我初次認識藤原浩這位人物，大概是在20年前。那時候是1991年，我是在雜誌《Hot Dog Press》某球鞋的文章中看到他訪問。不竟時間太久了，我大概忘記了那時到底是怎樣的一回事，但唯一我記起是當年身為高中生的我，就是從這篇文章初次發現球鞋的世界，同時激發起我的興趣。

我還能記起當時所受到的強大衝擊，那篇文章的內容是藤原浩10大最愛球鞋特集，到現在我還有這篇文章的副本，我大概可以告訴大家內容寫了甚麼、球鞋的圖解及把我當年閱讀時的感覺。

內容提及的第一雙鞋，是NIKE黑×白的COURT FORCE。當時藤原浩是這樣描述的：「因為我會穿上COURT FORCE去玩滑板，所以它們現在都已磨損了。現在市面已沒有COURT FORCE的存在，但我還是想買一雙新的。」COURT FORCE呀！磨損了比新的更好看！

來到排行第2位的是AIRWALK的ENIGMA，「我尤其喜歡藍及米黃的配色。」顏色的搭配十分好看，加上那殘舊的狀態實在太型了！接著繼續看下去跳到第4位的NIKE的藍×黑AIR JORDAN 1，「JORDAN系列最棒

My First Memory of Sneakers

The first time I came across Hiroshi Fujiwara was almost twenty years ago in 1991 when I read an article about Sneakers in the magazine Hot Dog Press. I could be wrong of course. I don't remember exactly as it was such a long time ago. The one thing I do remember was that as a high school student who had just discovered the world of sneakers, the particular article blew me away.

I remember exactly the impact it had on me. The article was a feature on Hiroshi Fujiwara's top 10 favourite sneakers. I have a copy of the magazine at hand, so let me give you a rough idea of how it read, providing some captions from the article and recapturing my thoughts at the time as I go.

The first shoe metioned was Nike's Court Force in black and white. "I used to wear my Court Forces for skate boarding, so they are all worn out now. They're no longer on the market, but I would love to get my hands on a new pair." Court Force! They look even better worn out than they would new!

Second was Airwalk Enigma. "I particularly like the blue and beige coloring." Definitely a great colour match! The clapped out condition of these shoes are

的便是第一代。由於某些原因,我對剛面世的第6代沒有太大興趣。」不是吧?我才剛買了第6代……我還不知道第1代曾經推出黑×藍的配色!太有型了……來到第9位的是NIKE的藍×黃DUNK,「這雙鞋是我去韓國時購買的,我實在太喜歡這個顏色的組合,所以我當時便買下他們所有的存貨!(笑)」我也很喜愛DUNK呀,尤其是那麼麼罕有的配色呢!他真的買下所有存貨嗎?他最後的一(笑)又是甚麼意思?他究竟是誰?

這就是初次觸動我對球鞋產生熱情的回憶,自此藤原浩這個名字在我心中便成為了有型、不可思議及擁有很多球鞋的人物。

20年後的今天,為了拍攝本書的Archive部分,我便到訪藤原浩的家及工作室,看到的是數量多得無法估計的球鞋,全部妥善存放於紙箱裏。當時我的第一個感覺是「藤原先生的東西很有條理,我也希望擁有大量這種箱子。」打開箱子,我的初次球鞋回憶便記起來了,便在心裡跟自己說:「我記得這些球鞋了,我熱愛球鞋的起源便是從這些收藏中開始的!」

說到這裡,我想我慢無目的談及個人回憶的部分都足夠了,但最後想說

的是在參與聯合編製《Personal Effects》(Hiroshi Fujiwara/Magazine House 2009)實在獲益良多,也啓發了我推出《Sneaker Tokyo》的靈感。還有,我十分慶幸可以跟Detz Matsuda先生一起編製《Personal Effects》,當我翻看前文提及的《Hot Dog Press》編輯人員名單時,他的名字竟然出現了,當時的他正好是訪問者及撰文的人!是命運嗎?還是事有湊巧而已?自我與球鞋產生關係直到今天的這20年間,我總是覺得我所參與的眾多企劃都是聯繫到不同人物的層面。

《SHOES MASTER》總編輯
榎本一生

really cool!! Moving on to Nike's Air Jordan 1 in blue and black in fourth place. "The first model in this series was the best. For some reason I'm not interested in the sixth model that's just gone on sale." What?! But I just bought the sixth model... I didn't know the first model came in blue and black!! So cool... In ninth place was Nike's Dunk in navy and yellow. "I bought these when I went to Korea. I really liked the coloring, so I bought all they had in stock. (laugh)" I love the Dunk! Such a rare colour. He bought out all the stock?! What is that (laugh) for in the end … Who is this guy?!?!

This was where my current passion for sneakers all started. Hiroshi Fujiwara's name was from that point engraved in the back of my mind as a cool, amazing guy who had lots of sneakers.

Twenty years later, to borrow the sneakers to photograph for the Archive section of this book, I visited Hiroshi Fujiwara's office and home, where he has an absurd amount of sneakers stored safely in tides of cardboard boxes. My initial thought was, "Hiroshi sure takes care of his stuff. I wish I had this many." But when I started opening boxes I recalled my first memory of sneakers, and thought to myself, "I remember these! My love of sneakers originates from this very collection."

I think I've talked aimlessly enough about my own personal memories now, but lastly I would just like to add that I also had the privilege of co-editing "Personal Effects" (Hiroshi Fujiwara/Magazine House 2009). If it were not for that book, it's likely this edition of Sneaker Tokyo would never have been conceived. On a further note, I was fortunate to be able to work with a guy called Detz Matsuda editing "Personal Effects", but when I looked at the staff credits for aforementioned "Hot Dog Press", his name appeared as interviewer and writer. Fate, or simply a coincidence? In any case, in the twenty years since I first developed a connection with sneakers, I have always felt that the various projects I come to work on are all connected on some organic level.

"SHOES MASTER" Chief Editor
Issey Enomoto

Sneaker Tokyo vol.2 "Hiroshi Fujiwara" ⌬

Edited & Written by SHOES MASTER magazine
www.shoesmaster.jp

TAIWAN

Producer
錢翠雯

Editor in Chief
黎玄才 "鬼塚"

Translation
丁廣貞
鄭雅云

Designer
楊宗豪

JAPAN

Producer
Shin Kawase [OFFICE KAWASE]

Editor
Issey Enomoto

Art Director
Koji Taniishi [GUTSON]

Designer
Tanny and Gomes. [GUTSON]

Design Coordinate
Shinichiro Komure [GUTSON]

Photographer
Osamu Matsuo [STUH]

Advertising Officer
Hirohito Iso

Advertising Staff
Yuji Kanazawa
Ryuji Kawakami

Director of Circulation
Kazuya Suzuki

Printing Director
Shinya Kinugawa

Translation
Catherine MacLellan
Midori Uchida

Special Thanks
Atsuyo Kitano

SNEAKER TOKYO

2010年6月8日

發行人/錢翠雯

很有文化股份有限公司

台北市松山區民生東路5段36巷4弄41號2樓

TEL：886-2-2762-3035

FAX：886-2-2762-4165

Marine Planning Co.,LTD

2-2-3 8F Jinboutyo Chiyoda-ku Tokyo Japan 〒 101-0064

MPC CO.,LTD

4-2 3F Shiraho Kanazawa-ku Yokohama Kanagawa japan 〒

236-0007

TEL：048-770-5464

FAX：045-770-5469

製版· 印刷

科憶印刷事業股份有限公司

ISBN978-4-89512-375-4

總經銷

聯華書報社